T0256064

RANDOMNESS IN EVOLUTION

RANDOMNESS IN EVOLUTION

John Tyler Bonner

PRINCETON UNIVERSITY PRESS

Princeton & Oxford

Published by Princeton University Press, 41 William Street,
Princeton, New Jersey 08540

In the United Kingdom: Princeton University Press,
6 Oxford Street, Woodstock, Oxfordshire OX20 1TW

press.princeton.edu

ISBN 978-0-691-15701-6

LIBRARY OF CONGRESS CATALOGING-IN-PUBLICATION DATA
Bonner, John Tyler.
Randomness in evolution / John Tyler Bonner.
    pages. cm
Includes bibliographical references and index.
ISBN 978-0-691-15701-6 (hardback : acid-free paper)
1. Variation (Biology) 2. Evolution (Biology) 3. Biodiversity.
  I. Title.
QH401.B66 2013
576.5′4—dc23   2012032006

British Library Cataloging-in-Publication Data is available

This book has been composed in Adobe Jensen Pro

Printed on acid-free paper. ∞

Printed in the United States of America

10 9 8 7 6 5 4 3

CONTENTS

ILLUSTRATIONS

PREFACE

Many biologists, and I am one of them, live two lives at the same time. In one they work with organisms from day to day in the laboratory, or in the field. This is what keeps them in touch with their subjects—the real world that they find so fascinating. The other life is a concern for the big picture: how it all fits together. No one gives a better example of this double life than Charles Darwin, with his long and excruciatingly detailed study of barnacles, on the one hand, and his breathtaking excursion into evolution, on the other. From a biologist's point of view there can be no bigger picture than the latter.

In my case the microscopic cellular slime molds have been my barnacles, and my mini journeys into looking at the broader picture have been put forth in essays over the years. Often they have been concerned with bringing forth a hidden phenomenon, or showing connections between different realms of biology that are much more closely related than convention would have it. I have been led to seek the big picture, one that ties together facts and ideas that are generally thought of as entirely separate. For instance, years ago I was concerned that the study of the development of organisms was limited to very few organisms, such as sea urchins and amphibians, yet the variety of organisms that develop is great. I felt that comparative studies of development would be essential for generating new insights into the venerable study of embryology and its relation to evolution. Later I was struck by the similarities between cell societies that make up a multicellular organism and animal societies, such as those of ants and other insects and mammalian societies as found in various species of monkey: the same forces make them tick. Then I became intrigued by the ways the essential communication between cells during development fell in neatly with the then new discoveries of that time of Nikolaas Tinbergen and Konrad Lorenz in the rising field of ethology; it was a different way of looking at the evolution of development.

My "big picture" kept changing. I soon realized the great importance of the concept of life cycles, and this led me to appreciate the importance of size. Life cycles of a multicellular organism (such as ourselves) start off as a minute fertilized egg and keeps getting larger and larger until maturity. This is as true for a small alga as it is for a blue whale. Furthermore, if one looks at the history of our planet, it is evident that the maximum size of animals and plants progressively increases over geologic time (a matter to be pursued further in the pages that follow). This size increase is fundamentally tied with an increase in complexity—what I have called the "size-complexity rule." And all during this series of generalizations that I have indulged in over the years, I have kept my eye firmly fixed on the cellular slime molds, my barnacles.

This book is a continuation of my quest for the big picture. I work with microorganisms and I ask the question: Are they affected by natural selection the same way as large, complex ones? It is conventional to think that selection acts the same way on organisms of all sizes, and I make the case that it might not be so. I was delighted to find that in a brief passage, Darwin also suggested the same idea in one of the later editions of his *On the Origin of Species*.

Margaree Harbour
Cape Breton, Nova Scotia

RANDOMNESS IN EVOLUTION

CHAPTER 1

Life and the Riddle of Randomness

As biologists our great aim is to find order in all the diversity and complexity of the living world. This is something we all strive for: it is what gives us a feeling of fulfillment and satisfaction. We seek the rules that underlie living phenomena to make ignorance and confusion turn into clarity and order. This is what Linnaeus did by finding a way to classify the vast numbers of different kinds of animals and plants: he brought order out of chaos. And this is what Mendel did with crossing his peas to reveal the basic rules of inheritance, a discovery of such great importance that it provided the basis for the majority of the advances in biology—which have been profound—in the twentieth, and now into the twenty-first century.

The great advance of Charles Darwin in discovering natural selection was another momentous step forward. It explained how organisms could evolve, how they, through successive generations, could become optimally adapted to their environment. There is continual competition between individuals, and the winners are the ones that are successful in producing offspring, thereby passing on the advantageous traits. All biologists today are so in harmony with this idea that it frames all our thoughts, so much so that it has a way of obscuring for us some important and peripheral factors that seem to be less worthy of our attention. But I think they are important, and this book is an attempt to put one of those factors before our eyes.

The most obvious one is randomness. There is something about this idea that is unsettling to many, no doubt because it goes directly against the more comfortable feeling of the order that we all seek. For this reason there is less written on randomness in evolution than on selection, although it is by no means totally absent, as we shall see. Compared to natural selection, it is no great enlightening principle, and therefore it is often relegated to a background noise that really is not doing anything. Natural selection carves out novelties that lead to evolution: randomness seems to go nowhere; it just shuffles things backwards and forwards. While there is some truth to this way of putting the matter, it is fundamentally wrong, as I plan to show in some detail. All of evolutionary change is built on a foundation of randomness. It provides the necessary material for natural selection which then does indeed bring forth the order our inner mind so actively craves.

More than that, we will see examples where randomness is literally put to use as a way of managing a key step in development of an organism (and also in animal societies). In evolution randomness can, in some special circumstances, directly produce order.

The part that randomness plays in evolution differs with the size of the organism. In fact, this is what alerted me to the subject. For many years I have worked with small cellular slime molds, and because of them I have been much concerned in the matter of how size influences both the development and the evolution of organisms. As I bore into this interesting matter I realized that most evolutionary biologists, following the tradition of Darwin, think in terms of large or at least complex plants or animals, and assume that microorganisms are no different despite their small size. There are those who work on the evolution of bacteria and often can distinguish between their characteristics and those of larger eukaryotic organisms, but the prokaryotic world is in some ways a specialized subject, although one of great interest.

An important point should be made right in the beginning. In this book I will be concerned with the variations in morphology. Bacteria, and other prokaryotes, have a very limited morphological variation, and therefore my argument here only concerns eukaryotes, whose cells contain a nucleus and come in a great variety of shapes. Those vast numbers of varieties involve both unicellular and multicellular forms.

Chance in Evolution

The role of chance in evolution has a venerable history, and there has been a recent renewed interest. It was recognized early in the history of genetics that mutations were random. More recently, and for many years, this randomness was understood at the molecular level where one of the bases in a DNA chain substituted for another. There have been many attempts to show that in some circumstances mutation might force change in a particular direction, but these experiments have not stood up with time; the idea that mutations are random has long been generally accepted. It should be noted that this fact has been used by skeptics to doubt Darwin and his natural selection: how can the complexity and the beauty of a bird or a flower be explained by a mechanism rooted in the chaos of randomness! But indeed it can, and we have more and more evidence that there are numerous aspects of evolution besides mutations that involve chance. There is a primaeval notion that one cannot produce order out of chaos despite the fact that it is a common phenomenon.

Not only is mutation random, but the genetic events involved in sexual reproduction are peppered with chance events. Since the egg and the sperm each have half as many chromosomes as the other cells of the body, and they arise with chromosome reshuffling during their formation, during meiosis, the genes any one gamete might possess will vary and the nature of this variation is a matter of chance. So stochastic or random events are very much involved in producing the genetic variation among individuals in a population; they help produce the variation that is the fodder for natural selection that makes evolution possible.

It was first pointed out by Sewall Wright, a pioneer of the surge in population genetics in the 1930s, that because of random events, such as the ones just described, the genetic makeup of a population could change simply because of those random events.[1] He called this stochastic evolutionary change "drift." It follows that this might be particularly important if a population was very small at one point in its history, for the variant genes it possessed would be the ones that remained when the population subsequently expanded. The whole genetic constitution of that population was founded on the genes that just happened by chance to be present when the population consisted of few individuals. It is obvious why such a bottleneck in population size would allow the chance event of "drift" to produce evolutionary change. This bottleneck phenomenon has also been called the "founder effect" because it might lead to the invention of a new species. It has also been called the "Adam and Eve effect," the ultimate in narrowness of a bottleneck. The important lesson from all this is that changes in the genetic constitution of a population can be determined by chance; chance plays an important role in evolution.

[1] For a review see V. Grant, 1977.

Another foray into the role of chance in evolution was made by C. E. Finch and T.B.L. Kirkwood.[2] They begin by taking note of the fact the duration of the life span of any animal is, within limits, entirely random. It is the result of an accumulation of accidents and not something that is consistent and controlled. Even genetically identical twins will show differences, not only in their life span, but in other characteristics as well. Furthermore, Finch and Kirkwood point out in great detail that many events during development are random and leave their imprint on the resulting adult. C. H. Waddington called this "developmental noise."[3]

One of the most important advocates for the role of random events in evolution is Michael Lynch.[4] He deplores the idea that natural selection accounts for everything and argues that random events play a significant role in evolutionary change, particularly in the evolution of complexity. His main concern is the evolution of the genome; he emphasizes that not only is the randomness of mutation key, but also the shuffling of the genome in recombination. He argues that, as in the drift of Sewall Wright, random molecular changes could give a directional push over time that does not involve natural selection.

The idea that chance plays a role in evolution has a venerable past and has been promoted by a number of individuals; it may therefore be considered an

[2] Finch and Kirkwood 2000.
[3] Waddington 1957.
[4] Lynch 2007a,b.

accepted notion. What might be new in my discussion will be the point that the effect of randomness differs for organisms of different sizes. This is the major argument I wish to pursue.

Size and Randomness

Evolution has been from small to big, from simple to complex. Besides this obvious point, there has been another neglected but equally important trend in the control—or suppression—of the effect of randomness. In microorganisms, random events are common, but with the increase in size and complexity there has been a corresponding decrease in the role of chance. So there are three phenomena: (1) the increase in size, (2) the increase in complexity, and (3) the decrease in the part played by randomness: all three go together during the course of evolution. And clearly they are interrelated.

While it is true that compared to small organisms, large organisms are protected to a considerable degree from the vagaries of chance, they nevertheless cling to some randomness; in fact, randomness is essential to their very existence. One need only remember that all novelty is founded on the generation of new genes, which arise from the directionless, random appearance of new mutations. So while large animals and plants, by a vast array of mechanisms, limit chance, they do so within a well-defined boundary: not too much, and not too little—something that is so admirably managed through the sexual system, which in turn is ruled, and created, by natural selection.

There is a danger in following this line of thought, i.e., that all organisms, over geological time, have pursued the same path; and if they have, why do any of those simple, primitive microorganisms still exist today?

Natural selection could not have it otherwise. The progression over many millions of years has not meant the elimination of simple, smaller organisms, but they also have been continuously maintained by selection or the lack of selection. So the role of selection in the great evolutionary history of life on Earth not only is responsible for the progressive changes in any one group of organisms, but also takes cognizance of the interdependence of organisms. The whole community is continuously under the stern eye of natural selection. To make the point by giving a simple-minded example, animals could not exist without plants, the ultimate suppliers of the energy for life through the process of photosynthesis. The role of any organism in the size-complexity-randomness spectrum exists because it fits in, and is part of the whole fabric of a community, all the result of natural selection. So, still existing today we have prokaryotes, protozoa, a great plethora of invertebrates, fungi, and lower plants, all of which joined the world eons ago. They have not been abandoned, and although they are continuously evolving, they remain within the group in which they originated. They are a permanent part of all evolution on Earth.

My friend V. Nanjundiah has pointed out to me that this very matter is discussed in Darwin's *On the Origin of Species* in an intriguing passage:

> Why have not the more highly developed forms everywhere supplanted and exterminated the lower? Lamarck, who believed in an innate and inevitable tendency towards perfection in all organic beings, seems to have felt this difficulty so strongly, that he was led to suppose that new and simple forms were continually being produced by spontaneous generation. I need hardly say that Science in her present state does not countenance the belief that living creatures are now ever produced from inorganic matter. . . . If it were no advantage, these forms would be left by natural selection unimproved or but little improved; and might remain for indefinite ages in their present little advanced condition. And geology tells us that some of the lowest forms, as the infusoria and rhizopods, have remained for an enormous period in nearly their present state.[5]

As we shall see, his answer to the question is not all that different from mine. In fact it looks very much as though he (and not Lamarck!) scooped me!

If we concentrate on the third major evolutionary trend—randomness and its progressively decreasing role—an unnoticed phenomenon is revealed.

[5] 3rd edition et seq., John Murray (1861) p. 135.

Eukaryotic microorganisms, in contrast to larger, more complex forms might have a greater prevalence of morphological variation, that is, they might be relatively untouched by natural selection (as Darwin suggests). This could possibly hold for numerous small organisms. As we will see, the great difficulty is that it cannot be proved, but it is a hypothesis that cannot be ruled out, either. This raises a very interesting point concerning the psychology of biologists.

Ever since I started to pursue these ideas some six years ago, I have been burdening respected friends and divers first-rate evolutionary biologists with early (and admittedly wanting) versions of this idea, and their criticisms have been enormously helpful. But as the process went on I began to realize that there was a bigger issue than correcting my bent sentences: the idea that biological diversity could be explained by something other than natural selection approaches heresy. The dogma, often stated explicitly, is that for any character in an organism for which its selective advantage may not be apparent, the safest assumption is that it is an adaptation and some day the reason for its selection will be revealed. This is so engrained in our thinking about evolution that the idea that stable morphological traits could be established during the course of evolution by chance is often dismissed without a thought. This is by no means universally true, and, as we have seen, there are some authors who, like me, bemoan the neglect of considering randomness's role in evolution. A good example was the publication of Stephen Jay Gould and Richard Lewontin's *Spandrels of San Marco*,[6] in which they argue that many of the justifications of calling something an adaptation is totally absent: they were *Just So Stories*, like those of Kipling. This was met with a deluge of counter-criticism, and it is clear to me that this outburst took place because the tradition of always assuming adaptation is so deeply embedded in our psyche. It is difficult to turn the page. I urge the reader to stay calm: I am not going to throw Darwin out with the bathwater.

One further point. The great difficulty in dealing with adaptations, or lack of adaptations, is that it involves a large share of speculation. It is an easy matter to make hypotheses and argue forcefully on either side. One way to deal with this problem is the use of mathematics. A reasonable mathematical model can often be very helpful, but all too often the model is substituted for reality. It may provide the perfect, satisfying solution, but in itself it usually involves assumptions, that is, hypotheses. Such a model certainly can be a very useful approach, but it may not contain the whole answer. At least in the best of circumstances it leads one in a helpful direction.

Using mathematics has been enormously effective in population biology, and indeed the great advances

---

[6] Gould and Lewontin 1979.

of R. A. Fisher, J.B.S. Haldane, and Sewall Wright are all the product of skillful mathematics. A while ago, when I was trying to convince a friend who is a population geneticist in the mathematical tradition that small organisms may be relatively unaffected by natural selection, he found the idea totally unacceptable. When I asked him if the difficulty was that he could not see how to put the matter mathematically, his immediate answer was yes. Mathematics in biology has tremendous power, but it cannot do everything.

As the upper size limit increases over geological time, there is a corresponding increase in the period of development that produces the mature morphology. This involves a great increase in the complexity and effectiveness of the mechanisms of control, and one result is a progressive stifling of the influence of randomness. The role played by randomness is significantly different between micro and macro organisms.

In large organisms there are many sequential steps in their extended development, each of which is under genetic control. If there is an unfavorable mutation in one of those steps, it will simply block any further development, and the embryo dies. This is what Lancelot Law Whyte called "internal selection."[7] The chances that any such mutation could be beneficial are extremely unlikely because all the steps that follow will be totally dependent upon

[7] Whyte 1965.

that previous step, so any change is very likely to be deleterious. Development has a built-in mechanism to eliminate undesirable random mutations. The larger the organism, the longer the sequence of developmental steps, and the greater the possibilities of internal selection.

In small organisms, with few developmental steps, one random change might not only affect the morphology of the organism, but often the whole organism. In this way it is possible to generate masses of different whole-organism forms, many of which might be unaffected by natural selection; one shape will do as well as another.

The key is the number of developmental steps: many, and randomness is suppressed; few, and the effect of randomness can come to the surface and bloom.

Sex

As I already indicated, randomness is the backbone of Darwinian evolution in the form of variation, in particular variation that is inherited. And the amount of such variation must be carefully controlled: too little means that selection does not have enough material to work with; too much means no change because of a glut of variants. And the sexual mechanism is a remarkably effective way of providing just the right amount of variation. Furthermore, this mechanism that makes evolution by natural selection possible is itself the product of selection. Having the right amount of variation leads to greater reproductive success. Sexuality is such an important element in natural selection that, not surprisingly, it is essentially ubiquitous. It has been pointed out by many that sexual reproduction, which yields fewer offspring per parent, is far more costly than asexual reproduction. But if that cost were not paid, there would be no evolution. Sex is the golden key to evolutionary progress. Evolution starts off on a foundation of randomness followed by the natural selection of a mechanism to control it.

In organisms that arose early in Earth history (invertebrates from protozoa to sponges, cnidarians and upwards; lower plants from algae to mosses and in between; and let us not forget fungi), there is a great variety of ways in which sex appears in their life histories. In the simpler forms, asexual cycles are often interspersed with sexual ones: the former are clones and generally have no variants and are present in a benign environment where they can multiply rapidly, while the latter characteristically appear in a changing environment where variation might produce some individuals that are more likely to be able to cope successfully with a change. At a later time in Earth history, when the larger and more complex animals and plants appear, there is no longer this switching back and forth of sexual and asexual phases: the asexual phase disappears almost completely. With a few exceptions, large animals and higher plants have lost the ability to have an asexual cycle.

This is the apex of control in the great sweep of organic evolution, but not everything is rigidly controlled; in fact, there is an underlying foundation of all of Darwinian evolution for organisms of all sizes, and that is a randomness. Without it there would be no evolution. Mutations are random, and without mutation there could be no change. So we see that while many of the stochastic processes found in small organisms have been diluted and to some degree silenced in the larger ones, the randomness of mutation is retained and is essential at all levels or stages of evolutionary progress.

In the pages that follow, the points made in this brief abstract will be greatly amplified. Chapter 2 begins with a description of the increase in size and complexity over geological time, from multicellularity (cell societies) to animal societies, with an attempt to understand the periods of little change with those of relatively rapid change. Next, in chapter 3, I will review how morphological randomness is dealt with at different size levels, beginning with eukaryotic microorganisms, where we see the greatest amount of morphological randomness in both aquatic and terrestrial forms. Chapter 4 will explain why randomness is curtailed in larger forms. Chapter 5 discusses how the sexual cycle also varies in a general way depending on the size and complexity of organisms. A partial reversal of my main contention that randomness is more prevalent in microorganisms is found in some smaller forms, for periodically in their life cycle some species suppress randomness by turning to asexual reproduction. Finally, I discuss two cases of great interest in chapter 6 where, in cell and insect societies, there is a small reversal, and randomness is brought back to the fore to play a key role in their respective developments.

CHAPTER 2

Time, Size, and Complexity

Since the argument of this book frames evolution in terms of size, it is important to look closely at that evolution of size. It is essential to understand how it occurred, and in particular why it occurred.

We are all brought up with the idea that life on Earth began over three billion years ago, and the first real organisms were prokaryotes. As life progressed since then, some organisms became larger while others remained small, giving us the incredible array of living creatures that exist today: from bacteria to whales, elephants, and giant sequoias, and all those plants and animals of intermediate sizes. There is general agreement that all those changes, and all those instances where there seem to have been no changes over great periods of time, are the result of natural selection. Let me now examine this evolution in more detail.

The evolution of size is the simplest and most obvious phenomenon, and clearly the range of sizes

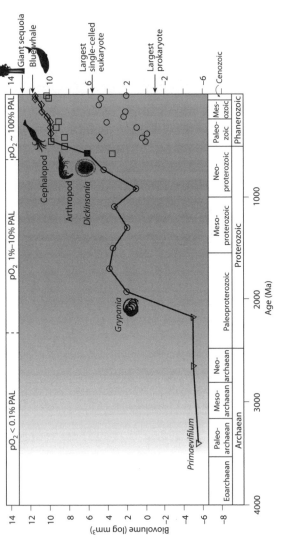

Fig. 1. Sizes of the largest organisms through Earth history. Size maxima are illustrated separately for prokaryotes (triangles), single-cell eukaryotes (circles), animals (squares), and plants (diamonds). Also indicated are the estimated oxygen levels. (From Payne et al. 2009)

has expanded steadily over geological time. For a number of years I have championed the idea that the maximum size of all kinds of organisms increased since the first single prokaryotic cells of few billion years ago. My point was based on very little data: it was to some degree a fuzzy (but true) general statement of the obvious. Fortunately, the matter has recently been examined carefully by J. L. Payne and his co-authors, who have searched the paleontological record to produce a truly empirical picture of this principle (fig. 1).[8] While the general idea that maximum size increases over time is supported, the details turn out to be of great interest.

In the first place, it is not the smooth curve I had imagined, but it is one that rises in three major steps. It is difficult to know exactly why the three spurts occurred and why there was a long period of almost a billion years between the first two, when maximum size remained roughly constant, although I will presently suggest a hypothetical explanation. As a general principle it is obvious that, among other possible factors, the maximum size will be limited by the mechanical structure of the organism. Therefore, if new designs of the body plan that have arisen by numerous mutations and other genetic changes are such that they can support greater size, and if there is, as we assume, an ever present selection for larger forms, then the size ceiling can be expanded upwards.

[8] Payne et al. 2009.

I have long argued that a major driving force for the evolution of complexity is the natural selection for size. If that selection happens to be for size increase, then there will be a mechanical limit as to how big the animal or plant can become. For instance, an ancestral fungus might be under selection for greater dispersal of its spores, and this selection would favor raising the fruiting body up into the air. Without a special supporting device it can only do this for very short heights. If chance mutations lead to a stiffened supporting tissue, then obviously it can respond to the selection for size increase, thereby providing more effective dispersal. And if the organism gets its energy by photosynthesis, reaching up towards the Sun and trying to outgrow other competing plants for that Sun is obviously desirable and is encouraged by the natural selection of vascular tissue. The new composition of its construction has allowed the evolution of size increase to proceed. So the advent of structural innovations can release the constraints and change will follow, in this case for successful bigger organisms.

Let me add parenthetically that increased complexity in evolution could also arise without a change in size; this is a point explicit in the writings of D. W. McShea.[9] To give one example, it is obvious that at any one size level, mutations and other genetic changes that increase efficiency might also increase complexity. However, this is different

[9] McShea 2005; McShea and Brandon 2010.

from a great sweeping trend as we see for the size-complexity evolution, although McShea argues on theoretical grounds that there might also be size-independent trends of increased complexity. And it should be pointed out that within any group of organisms, such as mammals, there can be a huge ranges of sizes, for example, from shrews to whales. So the generalization that getting bigger requires the invention of new parts and an increase in complexity is a very broad one. Among other things, the ancestors of shrews were large; they have undergone a selection for smaller size. To put the matter in a nutshell, one can get more complex without an increase in size, but one cannot have a significant increase in size without an increase in complexity.[10]

The invention of the eukaryotic cell and the invention of multicellularity were critical events that permitted size increase in the distant past. It is not clear from the fossil record exactly when these two crucial events occurred, so we can only make guesses based largely on what we know from small organisms that live today. The complicated event that gave rise to the eukaryotic cell, we presume, involved the association, or fusion, of prokaryotic elements. It is a major structural change; to indulge in a metaphor, it is like moving from a tent to a luxury hotel. One can imagine that such a major step made it a unique event, and that all eukaryotes are descendants of one such event. But of course that is wild speculation.

[10] For a further discussion of this point, see Bonner 2006.

How could this event have led to the ability to become larger? One big innovation found in eukaryotic cells is the invention of microtubules, which play a key role in affecting the shape of cells. Prokaryotes can only master simple rods or spheres with a few minor variations; eukaryotes have almost unlimited possibilities for varied shapes. These treasure troves of shapes have not in themselves led to a big increase in cell size, but they have increased the possibilities of constructing elaborate multicellular organisms. It is as though one made a variety of different kinds of building blocks to build houses of different designs.

We know from merely looking at the different kinds of simple multicellular organisms that exist today that multicellularity must have arisen at least a score of times. Even prokaryotes such as myxobacteria, cyanobacteria, and actinomycetes have multicellular species, and among eukaryotes they were invented independently by fungi, diatoms, protists, amoebae, and the algae and primitive metazoa. Among all those inventions only the latter two led ultimately to great size—trees and whales.

The first spurt in fig. 1 reached a peak early and then leveled off for almost a billion years. The spurt itself involved a considerable size increase over time and a number of organisms. Furthermore, this is an active field of research and new forms are being discovered constantly. One established example is *Grypania spiralis*, a fossil species of a somewhat enigmatic nature. There has been some discussion whether it is a prokaryote or a eukaryote, although the latter possibility is favored. Over some millions of years, the size of these baffling organisms increased, which might mean that at least some were multicellular.

This rise is followed by a long interval of almost a billion years, during which the maximum size remained pretty much the same. No doubt during this period organisms of different degrees of complexity did arise. As I just pointed out, there can be selection for increased complexity that is independent of changes in size; the added complexity will in itself have arisen because of its selective advantages. These advantages could provide an increase in efficiency that would lead to competitive success. My only claim here is that if, through selection, the size of the descendants of an organism increases, some changes in complexity will be required before it can happen. It cannot proceed in this upward trend without the necessary structural changes to sustain it, which in turn will arise by mutation and natural selection.

So, during that billion years numerous inventions that led to increased complexity undoubtedly occurred independently and at different times. Most of those early cases of multicellularity remained small; they did not significantly challenge the upper size barrier. However, there were two successful ones that made it possible to achieve great size; they are the inventions that gave rise to animals and plants (and large fungi with their stiff cell walls). Here I would like to make a bold hypothesis. The reason for the roughly billion-year period where there is no overall increase in size, as is shown in fig. 1, is that it took that long to invent multicellular constructions that would be capable of further size increase—namely, that of plants, fungi, and animals. Certainly all the other independent inventions of multicellularity referred to above did not evolve into any kind of organisms of large size, possibly because the way they were built made it physically impossible. But finally, after millions of years, two magic formulas emerged that allowed the rise of larger animals and plants (and, to a more modest degree, fungi).

The beginning of the evolutionary success of these two groups must be sought at the cellular level. In the case of plants (and fungi), the invention of stiff cell walls clearly provided strength. In terrestrial forms, strong cell walls allowed them to stick up into the air. In aquatic algae, where cell walls first appeared, stiff walls allowed them to achieve great size, as in marine brown algae such as kelp and other forms. The cell walls not only hold the cells together to provide some structure to the multicellular mass, but give it strength as well.

In the beginning of the animal line, motility seems to be the ruling activity. Their ancestor cells were either amoeboid or flagellated to provide external mobility to capture food or escape predators. This led to the invention of muscle and, to hold it, bone and other kinds of rigid skeleton, making a further increase in size possible.

The basis for the great success of animals, plants, and fungi to become larger lies in a fortuitous combination of cellular processes at the beginning of their evolution. All the other experiments in multicellularity did not possess the magic formula and evolved no higher than colonies of modest complexity and size.

The Problems of Getting Larger

There are two basic requirements for size increase in animals. One of them is the same as that of plants: to support increased mass there must be some sort of supporting structure, as found primarily in skeletons, exo- and endo-, that have been reinvented more than once in the evolution of animals. A fundamental characteristic of animals is that they can move; they have been selected for locomotion so that, among other things, they can capture food and escape from predators, activities that are not needed in photosynthetic plants (with the partial exception of some carnivorous plants).

The other basic requirement is that key molecules be allowed to enter all parts of the large body so that metabolism can take place. This process involves diffusion. Most animals are aerobic, that is, they require oxygen to burn their food; it is the key molecule that provides, through combustion, the energy for locomotion and all those other living activities. For metabolism to occur, sufficient oxygen is needed and it can diffuse about 1 millimeter into an organism. So any small, nonphotosynthetic organism, such as a single-cell protozoan, can manage very nicely without any special devices. But with the most modest increase in size the problem of the immediate availability of oxygen to all parts of an organism becomes a crucial issue. One solution is for the surface of the organism to become quite flat, or highly convoluted to increase the surface-to-volume ratio. However, there are obviously severe limits to how much bigger one could become if those were the only mechanisms available to get oxygen into the inner tissues.

The best way to understand how larger animals deal with this problem is to go to a mammal, such as ourselves, and see how we manage. Just for gas exchange, taking in oxygen and getting rid of the toxic carbon dioxide, both involved in the combustion that takes place within our cells, we have two major and extremely complex systems that have arisen during the course of evolution, making it possible for our deep tissues to survive. These are the circulatory and the respiratory systems. The complexity of both of these is quite extraordinary. The circulatory system requires a pump—the heart—and a vast network of vessels that reach all the cells, even those in the deepest tissues. The blood carries the oxygen (and carts the carbon dioxide away) by means of haemoglobin, an amazingly clever and complex protein. The blood also carries the food to the inner cells primarily in the form of small molecules such a sugars and amino acids.

There is more than one device for getting oxygen to the cells. There has to be a way of getting the oxygen from the air to the blood and its haemoglobin. For this we have elaborate lungs that take in the air by means of the muscles (and nerves) that control breathing, and they are so constructed that blood capillaries come in close contact with tiny air sacs bearing the incoming oxygen.

The big animal problem is not yet solved. There has to be a mechanism to get the food into the cells. For this we have a gut into which food is taken in through the mouth and immediately attacked by digestive enzymes that break down large molecules such as proteins and polysaccharides into small amino acids and sugars that can move through the gut wall and enter the blood stream, where they are taken to all the cells in the body. They are the fuel for the cells, and their combustion involving oxygen provides the energy for all living activities.

Payne et al. point out that during the course of the evolution of life there is also an evolution of the amount of oxygen in the atmosphere. As can be seen in fig. 1, they have indicated three periods of successive oxygen increase that correlate with the different periods of maximum size that they describe. One immediately wants to know if this oxygen increase is a cause of the maximum size increases. It seems to me that those increases in oxygen made the size increases possible; they did not cause them, but enabled them.

Using a mammal as an example of how oxygen and other substances can reach the deep tissues makes the point that special devices are necessary, but they obviously did not arise full blown. As the size of animals increased over time, an increase in the sophistication of devices for handling the diffusion problem arose. If we turn to smaller multicellular animals we find many different and often less elaborate solutions to the same problem. For instance, small nematode worms and rotifers have a very simple gut—merely a short straight tube—very different from our exceedingly long and highly convoluted gut with its associated glands that produce digestive enzymes, and special structures to increase the surface area to accommodate the intake of all the food to keep our great bulk nourished. Nematodes and rotifers are small enough so that diffusion can occur without any special devices.

On the other hand, insects, an incredibly successful group, are for the most part sufficiently big so that a special mechanism for gas exchange is required. They have devised a system that is entirely different from that of mammals. Instead of lungs their tissues are riddled with minute tracheae, air tubes that have numerous openings to the air along the side of their bodies. The tubes penetrate deep into the tissues so that they reach the inner cells. Furthermore, they have a crude sort of breathing in that any movement of their abdomen helps move the air in and out of the tubes.

The necessity for special structures to allow for size increase is a fundamental feature for all animals, but the solutions are not necessarily the same. Different structures arose independently by natural selection; they are examples of convergent evolution. In the above cases they permitted the evolution of size increase, but of course the size of insects comes far short of elephants and whales. The fact that insects are limited by their tracheal system is only one of the reasons they have never achieved great size; another is that they have an exoskeleton that lacks the structural possibilities for increased size that is found in a body with an endoskeleton. All this shows again how particular innovations in construction set a limit on how large an organism's descendants can become.

Complexity

These sophisticated solutions in coping with size increase obviously did not arise suddenly, but had a long and gradual evolution. One way to see this clearly comes from a study of Valentine and his co-workers,[11] who estimated the maximum number of cell types in animal fossils from different geological epochs (fig. 2). As is clear, with time there is a steady rise from a handful of cell types to over two hundred: from the first multicellular invertebrates

[11] Valentine et al. 1944.

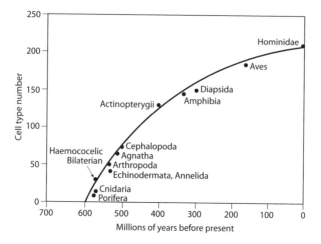

Fig. 2. Estimated numbers of cell types of early members of various animal groups. Only members of the groups that are believed to have been near the upper limit of cell type numbers when they originated are included. (From Valentine et al. 1994)

to mammals. They point out that for animals a new cell type arises on average about once every three million years.

Their curve is for the same time span as the third spurt shown in fig. 1. This means that the increase in size and the increase in complexity arose together, consistent with my contention that size is a prime mover. There is always selection for size change, and if it is for size increase, it must necessarily be accompanied by an increase in complexity for the very reasons mentioned above.

Measures of complexity are fraught with problems. In the first place, there is no easy definition. Complexity is generally thought to involve the interactions, the interconnections between parts. The human brain, which is a perfect example of a complex system, not only has a vast number of neurons, but an even greater multitude of interconnections by means of synapses that link the neurons. Furthermore, all the neurons are not the same and some have different functions from others; there is a division of labor. Therefore we can define the complexity of the brain in terms of these three components. And if we were asked to quantify brain complexity in any precise way, it is obvious that we are faced with a problem that is perhaps impossible to solve with exactitude.

I use this example to make a point. For our purposes here we do not need a precise measure of complexity; a very approximate one is quite adequate. All we need is some sort of measure that one group is more or less complex than another.

If we are concerned with multicellular organisms, the common method is to consider the number of cell types, which are an index of the division of labor. This says nothing of the interconnections between the cell types; it just assumes—quite reasonably—that the more cell types, the more interconnections. Even an estimate of the number of cell types is exceedingly inaccurate, but it is sufficient to arrange different organisms in a rough hierarchy (as Valentine

et al. have done in fig. 2). One of the reasons that
the exact number of cell types in any one multicellular organism is so difficult to determine is that it
depends on human judgment. The decision of how
different two kinds of cells have to be to be considered distinct cell types may not be the same for different observers. But this kind of inaccuracy will not
affect the general observation that one group has
more or fewer cell types than another.

The next question is, how can we compare the
complexity of single cells with that of multicellular
animals and plants. It has always seemed to me intu-
itively that any individual cell is exceedingly complex
and the added complexity acquired with multicellularity is modest by comparison. Eukaryotic cells
contain an elaborate division of labor: think of all
the internal structures with discrete functions: the
chromosomes, the nucleus, the Golgi apparatus, the
plasma membrane, the cytoskeleton, the mitochon-
dria, to mention some major ones. And each of these
has, to varying degrees, a further division of labor
among its parts.

How can one number those parts, as was possible
for cell types? As McShea[12] has pointed out, each
level involves a separate index of complexity. One
way might be to consider the level of cell complexity in terms of the kinds of molecules or organelles,
which would be levels lower than cell types. This

[12] McShea 2002.

would reflect the degree of division of labor within the cell. Perhaps there is no point of actually trying to put numbers to those molecule types, but the number would be very large, very much larger than the two hundred or so cell types in a mammal. McShea makes the interesting point that if one goes from one level to another, such as cells to multicells, the lower-level cells will lose some of their complexity and have a reduction of organelles when they become part of a multicellular organism.

It might be helpful to compare eukaryotes with their ancestral prokaryotes. Clearly the latter have far fewer organelles and a much simplified interior structure, and one can say with confidence that the prokaryotic cell is less complex—that is, it has fewer molecule types—than its descendant, the eukaryotic cell. Yet both of them, in their own ways, are far more complex than the community of cells that make up any multicellular organism.

Stimulated by the graphs of Payne et al. (fig. 1) and Valentine et al. (fig. 2) I have drawn a graph that incorporates theirs and adds the points about complexity just discussed (fig. 3). Let me hasten to say that much of the figure is totally imaginary because it is not based on specific measurements of complexity, but on generalizations just made that are hopefully reasonable. It assumes that unicellular prokaryotes are less complex than eukaryotic ones, and that the addition of multicellularity has produced a relatively modest further increase in complexity.

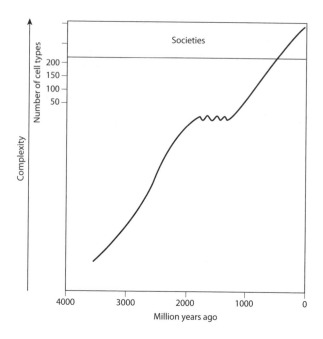

Fig. 3. An imaginary graph roughly estimating the increase of complexity of all organisms and organism groups over geological time. As McShea (2002) has pointed out, the measurement of complexity is different for each level and therefore this graph commits the sin of mixing oranges, pears, and apples. The complexity of cells can be measured in terms of the number of different kinds of molecules, or the various organelles, while multicellular forms are most easily characterized by the number of cell types. On the other hand, animal societies are gauged by the number of castes. Even though each level is different, they are nevertheless additive, albeit in a rough way.

Despite the hypothetical nature of some parts of this graph, it has a very important message that bears on many of the points that will be touched upon in this book. As can plainly be seen from the time scale on the figure, many millions of years are involved. This means that prokaryotes have been around for an incredibly long time, as have eukaryotes. Even the inventions of multicellularity go back millions of years. Finally, the most recent spurt of multicellular complexity increase has an ancient history and spans innumerable generations.

As just noted, the estimate made by Valentine and his colleagues for a new cell type to be invented is, on the average, about 3 million years. For an intermediate-size organism in which a generation takes about ten years, there would be 300,000 generations in 3 million years. Presumably this would be ample time for a group among the randomly arising new mutations to organize new genetic arrangements to carve out a new cell type. Such estimates are very inaccurate and probably not universal. For instance, M. Herron and R. Michod have done a molecular phylogenetic study of *Volvox* and its ancestors and find that they took about 200 million years to produce a second new cell type, and have done so independently in at least three separate lineages.[13] Because their generation time is very short, such a time span

---

[13] Herron and Michod 2008. I thank Matthew Herron for this information.

means an incredible number of generations. This makes an obvious point: there is plenty of time for natural selection to produce major innovations.

Let us now compare the complexity of multicellular animals with that of animal societies. I have simply added the animal-society complexity to the internal complexity of the individual animals (fig. 3). I have done this in a manner that grossly oversimplifies the differences but makes a correct very general point. It is totally reasonable to add the complexity of the society to the complexity of the individuals, even though the units of complexity used are quite different. I have consciously committed the error of adding apples to oranges.

I should mention some details that are not in fig. 3. In the first place, insects have about half the number of cell types of mammals, so even though they are capable of very large and complex societies, their total level of complexity—their addition of apples and oranges—would be less than that of a mammal. Furthermore, in both groups the size and the degree of complexity vary enormously. Another matter is when they became social. There is some evidence that social insects first appeared in the Cretaceous, very roughly 100 million years ago, but we have no such firm information for mammal societies—or any other vertebrate societies, for that matter. So all we can do is guess in very general terms, and it is likely that mammals' social origin is more recent than that of insects.

One difficulty is that the complexity of human societies is off the chart compared to any other animal
societies. Therefore I have not included them in fig.
3, which is confined to nonhuman animals.

If one compares the complexity of cell societies
with that of animal societies, the latter are far less
complex. This is true on any level. The number of
signals, beginning with chemical signals, are far
fewer. The division of labor, even in complex insect
societies, is very reduced: compare four castes of a
huge ant colony with the hundred or so cell types in
an individual ant.

This raises the question of why such a difference
exists. One possibility has to do with the fact that
the units in a society are generally mobile and not
attached to one another. This is related to the fact
that the genetic control of activities is not as central-
ized in animal societies; the whole control structure
is relatively loose and spread about. Another possibility is that cell societies have had much more time
to organize: they have been evolving for 1,000 million years, while animal societies have evolved only
for 100 million, a tenfold difference. Does this mean
that in the next 10 million years animal societies will
match cell societies in their complexity? This seems
rather unlikely, but the question makes us pause.

If one now thinks of these changes in terms of
mutation and the rearrangement of the relation-
ship of existing genes, the longer the time span the
greater the number of mutations and the greater

the number of new patterns of gene regulation. The initial step towards complexity is, needless to say, mutation. However, in prokaryotes, radical morphological inventions are rare, as, for example, the invention of eukaryotes. Most of the novelties due to mutation are on a biochemical level and give rise to changes that occur continuously, fine-tuning as the organisms' environment changes. This is the norm for prokaryotes: over billions of years they have not become significantly more complex in their morphology. Their niches tend to be biochemical and the chemical environment is what guides their evolutionary progress.

Here I am concerned with those mutations and genetic changes in eukaryotes that have led to the rising trend in complexity over time that we see in fig. 3. As I will argue, every innovation first started as a single mutation (or perhaps a small group of mutations) that produced a selectively advantageous phenotype. It was followed by many subsequent mutations and interactions among the genes that were either advantageous or harmful. So as evolutionary time passes, not only can the phenotype become more complex, but so can the corresponding genotype. As we shall see, there are important consequences of this fact. The organisms that exist today, whether they be an elephant or a flea, contain an enormously large number of genes and gene complexes that they have accumulated over millions of years and became adept at fine-tuning the phenotype. These genes and their arrangements have been favored by natural selection, while a host of other mutated genetic activity has been cast aside.

Animal Societies

There is another aspect of the evolution of complexity that I have indicated on my speculative graph. Not only does it show the progression from prokaryotes to eukaryotes, and from unicells to multicells, but also the leap from single organisms to animal societies. This too involves an increase in size: an ant colony or a baboon group are obviously collectively much larger than an individual ant or baboon. Furthermore, the groups are more complex, for the complexity of the social group is added to the complexity of the individual organism. It will involve additional division of labor and new systems of communication to bring the society to a higher level of complexity. And let me repeat the point that over the great geological time span this progression of increased complexity goes hand-in-hand with an increase in size. In any major way the two are inseparable. (This part of fig. 3 is grossly oversimplified, as explained in the caption.)

This chapter has been a very brief outline of the grand progression of evolution since life first appeared on the surface of the Earth. Now we will look at the decreasing role of randomness in the trajectory from small to large organisms.

CHAPTER 3

Small Organisms and Neutral Morphologies

The possibility is raised here that microorganisms might in some circumstances have *neutral morphologies*. By that I mean that there are significant differences between individual morphologies, but natural selection is blind to them. I will suggest candidates for this unconventional condition for both aquatic and terrestrial forms. In particular, I will be considering eukaryotic microorganisms, and not bacteria and archaea. The concern here is with morphology and the degree to which it is adaptive, and bacteria and archaea have very limited morphologies; their variation exists mostly in the form of multitudinous biochemical differences.

Let me preface the discussion by pointing out that mutations that affect morphology are far more likely to have an immediate effect on that morphology in small organisms than in large ones. Small organisms have short periods of development and any mutation, or genetic change, will appear directly in the phenotype, and in doing so will likely affect the morphology of the whole organism. In contrast, the development of large, complex organisms is a long and extended process, and viable mutations, or genetic changes, usually only affect a small part of the whole. So small organisms have a far greater possibility of a large range of radically different morphologies.[14]

What I propose here is the possibility that in some cases the morphology of eukaryotic microorganisms might be unaffected, or weakly affected, by natural selection. They are *neutral morphologies*. If many species of different morphologies coexist, each apparently being equally successful, then their morphologies might be neutral. Or if an organism remains unchanged morphologically through successive generations without being altered by natural selection, it might possibly not change because it is neutral. Neutral morphologies must be distinguished from unchanging morphologies due to stabilizing selection, where deviations from the norm are selected against: the morphological constancy over many generations is not a matter of chance, but of continuous selection.

[14] Let me say right in the beginning of this chapter that the morphological differences that are of concern here are variations between species, not variations within a species. This is a very general statement because it is not always easy, especially among microorganisms, to be certain what exactly is a species.

A Neutral-Morphology Theory of Evolution

Thus we have a neutral-morphology theory of evolution, where a variety of morphologies are equally successful in a particular environment. This makes an interesting contrast to the neutral-gene theory of Motoo Kimura.[15] In the former, for one reason or another, natural selection fails to discriminate among phenotype morphologies, each of which has a distinctive genotype; in the latter, selection fails to discriminate among genotypes that all could have the same phenotype.

Neutral morphology should be considered a null hypothesis, that is, a hypothesis that is assumed—by default—to be true until it is proved to be incorrect. The idea of a null hypothesis was put forward by R. A. Fisher many years ago as a statistical exercise: null hypothesis could be tested and proved or disproved by comparing it with another set of data to determine if it was statistically different or the same. I do not see how such a test could be possible for neutral morphologies. Here, to test the null hypothesis one would have to show that any one presumed neutral morphology is indeed adaptive, which would automatically remove it from its "null" condition.

As we have seen, over the course of evolution there has been an increase in the maximum size of animals and plants, including a progressive extension of the period of development that involves an

[15] Kimura 1983.

increase in complexity. With this increase in size, the possibility of neutral morphologies decreases. Once the eukaryotic cell was invented, there is every reason to believe that initially all the early species were unicellular and that multicellularity arose subsequently. As multicellular organisms increased in size over geological time, they and their development became increasingly complex. This means that neutral morphologies are a very ancient phenomenon and persist today in microorganisms.

Aquatic Habitats: Radiolaria, Foraminifera, and Diatoms

In the latter part of the nineteenth century the *HMS Challenger* made a voyage around the oceans of the world and collected a vast quantity of fauna and flora. From this richness they asked various biologists to examine particular groups and Ernst Haeckel was given the radiolarians. He described more than four thousand species (although today some fifty thousand species are known). The skeletons, consisting mostly of silica, vary fantastically in delicacy and form, as beautifully illustrated in the magnificent drawings of Haeckel (fig. 4 a & b).

The idea of phenotypic neutrality in the morphology of radiolaria and foraminifera stems directly from D'Arcy Thompson's discussion of their shells in his *On Growth and Form*.[16] There are two motivations

---

[16] Thompson 1917 et seq.

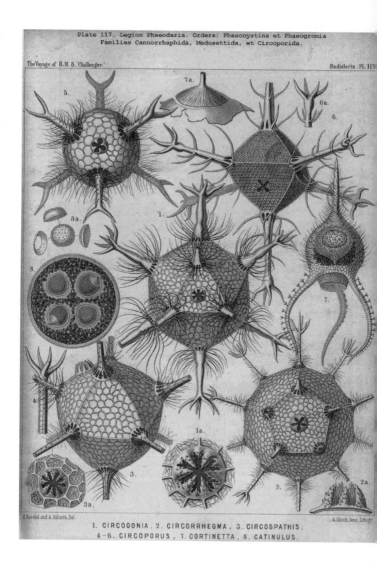

Fig. 4, a and b. Two plates from Ernst Haeckel's study of the
Radiolaria collected on the voyage of the HMS *Challenger*
(monograph published in 1887). He described over 4000

The Voyage of H.M.S. 'Challenger.'

Radiolaria Pl. 140

E.Haeckel and A.Giltsch del.

E.Giltsch Jena Lith.sc.

1-3. DIPLOCONUS. 4-8. DIPLOCOLPUS. 9-12. HEXACONUS.
13. 14. COLEASPIS. 15. 16. HEXONASPIS.

species of fantastically diverse shapes and these two plates
give one a sense of the extraordinary variety.

behind his discussion: he was particularly interested in the physical forces that produce shape, and he was quite down on the Darwinian concept of natural selection. Within this frame he gives examples of what I would consider possible candidates for neutral morphologies that are not being buffeted by natural selection. And indeed physical forces do play a role in the shaping of all biological forms, as he so effectively argued.

He goes on to say that not only are these beautiful creatures living forms, but many species have remained much the same through their geological history. "From the Cambrian age downward, the families and even genera appear identical with those now living." This is consistent with the idea that they might be neutral phenotypes, although that does not preclude the possibility of some natural selection. For instance, the variety of polar forms of radiolaria are in some ways different from those in the tropical oceans, indicating some adaptation has taken place; the forms are nearly neutral.

D'Arcy Thompson applies the same argument to foraminifera and their calcareous spiral shells, a group that consists of some 270,000 species. The structure of their skeletons nowhere near matches the intricate beauty of radiolarians, but he still uses them for his argument against natural selection. This is not quite so obvious for the foraminifera because they, too, have left a long fossil history and show an evolution from simpler forms to larger, more complex ones in later eras. In his day, others argued that these changes were all adaptive and the result of selection to produce "the survival of the fittest," but D'Arcy Thompson would have none of this. In particular, others argued that the more elaborate shells led to greater strength, an argument that Thompson considered an absurdity for purely physical reasons. As in radiolaria, there is evidence of some selection, not only in changes over time, but the forms that lie deep in the ocean differ from those at higher levels.

The very same argument for neutrality also applies to diatoms, unicellular photosynthetic organisms found both in salt and fresh water. They are encased in silica shells that are like a box, with a bottom and a lid. The variety of their shapes is also quite fantastic: it is estimated that there are about 100,000 species, each one with a differently sculptured shell (fig. 5).

So here are three unrelated groups of organisms that have independently shown an incredible variety of forms that not only exist today, but many have remained unchanged, or little changed, for millions of years. It is difficult for me to imagine that each of those many, many thousands of species, each with its distinctive sculptured shell, is maintained by a specific act of natural selection. It has even been suggested to me that each shape is a selective response of a specific predator, or a specific niche, but if so, where are those thousands of predators and niches?[17]

[17] See Nee and Stone 2003 for a pertinent discussion of neutrality in plankton.

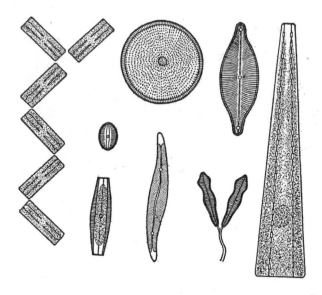

Fig. 5. Some of the common forms of diatoms. (From W. H. Brown. *The Plant Kingdom*, Ginn, 1935)

These three groups of organisms represent three good possibilities that are close to having morphologies that are neutral. It cannot be said that they are in fact neutral, but only that neutrality is a possible explanation for their great morphological diversity. Neutrality is also consistent with the fact that some species have remained relatively unchanged over great stretches of geological time, which could be the result of the absence of selection. This is a matter to which I will return shortly in the discussion of randomness in cellular slime molds.

"Everything is everywhere"

Neutrality among microorganisms has been approached from a different point of view. There has been considerable discussion about the distribution of aquatic eukaryotic microbes: is "everything everywhere," or are they, like larger forms, mostly in niches and as a result restricted in their distribution compared to the species that are truly cosmopolitan? B. J. Finlay and T. Fenchel have been stressing that a large number of ciliates and other aquatic, unicellular eukaryotes are cosmopolitan: the same morphological species are found all over the globe.[18] They consider that the main contributing factor to this phenomenon is their small size along with their vast numbers and therefore their high rate of dispersal. They back up their contention of everything being everywhere with much convincing data from these aquatic environments.

Others argue that cosmopolitanism is often not the case, and that particular species are found only in special niches. For instance, it has been maintained that oceans are often not just one great stirred caldron, but have layers and currents that can provide separate niches. This matter was examined by Laura Katz and her colleagues, who showed that sequestered bodies of water along coastal marshes and pools tended to have more species of ciliate protozoa compared to the cosmopolitan species of the

[18] Finley and Fenchel 2004 which includes references to their previous publications.

open ocean.[19] All is resolved if one thinks of a continuum between total morphological neutrality with no selection at one end, with strong selection at the other, and all different degrees in between.

Let me add a parenthetic note here. The profusion of species among these microorganisms appears to border on the matter of species diversity of macroorganisms and its causes, a current topic of great concern. The diversity theories either stress the occupation of diverse niches, or the effect of chance, either of which (or both) might play a role. So far this discussion has been almost entirely about large organisms, but there is a growing interest in examining the case for microorganisms.[20]

Terrestrial Habitats

There is a big difference between the life of aquatic and terrestrial microorganisms. For organisms that exist in the ocean or in other large bodies of water it is harder to find niches that might separate incipient species. No doubt they exist, but they are ill-defined compared to what one finds on land, and on land they are more plentiful for large organisms than small ones. Let us examine this proposition more carefully.

The whole subject of phenotype neutrality came to mind as a result of recent studies on cellular slime molds in their natural habitat, namely the soil. They

[19] Katz et al. 2005.
[20] E.g., Nee and Stone 2003.

are not unicellular and usually consist of two cell types: spores and stalk cells. They are small and are therefore possible candidates for neutral morphologies. The soil environment is very different from water: it is not in constant motion but rigid and structured. This means that many of its life strategies are different, including the basic mechanisms for dispersal, so important for survival. It also means that one can observe microecology in a manner quite impossible in the ocean.

A number of features of the life cycle of cellular slime molds bear on the arguments to follow. Let me remind the reader that unlike most organisms, they feed in their unicellular phase. Therefore if amoebae find a patch of bacteria they will consume it, and with the stimulus of starvation they will aggregate to commence their multicellular stage, which culminates in a small fruiting body: a stalk supporting a spore mass (or masses, depending on the species). This fruiting body is the amoebae's prime agent for spore dispersal: they stick the spores onto passing crawling beasts on the surface of the soil. The spores need to reach another patch of bacteria for continued survival.

That different morphological phenotypes do not seem to be discriminated by natural selection is one possible explanation as to why cellular slime molds of different morphologies can coexist in one small plot of soil. To take an obvious example, it is common to find a species of *Dictyostelium* flourishing side by side with one of the two common species of *Polysphondylium*, yet they differ radically in their morphology. *Dictyostelium* generally has a simple stalk crowned by a single globe of spores (a sorus), while *Polysphondylium* not only has a terminal sorus, but a series of evenly spaced whorls consisting of tiny stalks—like the spokes of a wagon wheel—each with a small terminal sorus (fig. 6). Since both kinds of fruiting bodies are structures to facilitate spore dispersal, what could be the advantage of one of these shapes over the other? Both might be equally effective in spreading spores by passing insects and other motile soil invertebrates. This is a hypothesis that cannot be ruled out.

An even more potent argument that all are equally effective as dispersal structures is that two of the more common (and cosmopolitan) species, *D. mucoroides* and *P. pallidum*. They are very often found close to one another in the soil all over the globe— maybe because their morphologies are comparatively neutral and one does as well as the other. They appear to be unaffected by one another's presence. Both are initially the result of selection as effective structures for spore dispersal—but the differences in their body plan make them good candidates for neutral morphologies. I am quite sensitive to the fact that there are ways in which natural selection could account for the facts just outlined, and there is no way of ruling them out, but I find myself surprised that neutrality is never even considered as a possibility. Friends have suggested numerous ways natural selection could account for the phenomena I

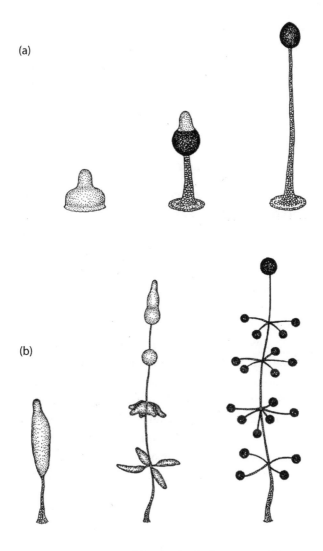

Fig. 6. A comparison of fruiting body formation of Dictyostelium (a) with that of Polysphondylium (b). (From Grell
1973; drawing by G. Gerisch)

have just described, but like mine, they are all *just so stories*. My *just so story* is that no selection is involved and they are neutral phenotypes. No doubt the day will come when we will be able to test the alternatives, but for the present all the arguments stand as possible hypotheses.

Niches and Size

There is another related matter of great interest concerning niches. Larger organisms have more niches available than small ones. There are a number of striking examples of adaptive radiation, such as the spread of honeycreepers in Hawaii and the finches in the Galapagos. In the former the ancestral bird—also a finch—came to Hawaii some 4 to 2.5 million years ago and radiated to fifty-six species of radical diversity, half of which are still found there today. The ancestral Darwin's finch arrived in the Galapagos some 2.3 million years ago and evolved into fifteen species. The kind of niche we are concerned with here is the available food. Hawaii was rich in a variety of foods, from the nectar of differently shaped flowers to various insects, fruits, seeds, and even snails. There has been strong selection for different shapes of their beaks. In the Galapagos the menu includes insects, blood from the backs of boobies, those large sea birds, and in particular seeds of different sizes. After a wet period, small seeds are abundant, and large ones in times of drought. In both cases these two food niches have resulted in the selection of different-shaped bills: massive ones for big seeds and small ones for small seeds.

Let us now consider the much smaller cellular slime molds. What is their food niche? The only organisms they can eat are minute bacteria (or very small species of yeast) that are engulfed by the separate feeding amoebae; in other words, they are restricted to essentially one food niche. This possibly explains why they have a very limited repertoire of morphologies despite the fact that they have been around for maybe 400 million years. The shape of the fruiting bodies has nothing to do with feeding, but only with dispersal.

Two Size Levels

In one stage in their life cycle the cellular slime molds are unicellular: the vegetative stage when the feeding amoebae roam around separately; at a later stage, the amoebae come together to form a multicellular organism.

E. G. Horn looked into the problem of how different species of slime molds could coexist in the soil, addressing the venerable ecological question of how organisms avoid competitive exclusion.[21] He isolated four species of cellular slime molds and a variety of bacteria from a small area in a local woodland. He then tested the ability of the slime molds to eat streaks of the different bacteria on agar. While

[21] Horn 1971.

all four could eat all of the different bacteria, they ate some more rapidly than others. He concluded that the amoebae of the slime mold species were able to cohabit in the same area because they had somewhat separate food niches and thereby avoided some competition and can coexist.

In the stage of the life cycle where the amoebae are separate and independent, we see the possibility for a standard selectionist explanation for coexistence. This might account for different species inhabiting the same plot of soil. Now we ask the question: What are the selective forces acting on the different fruiting body morphologies such as the presence or absence of whorls? Here we are faced with the possibility that they could be neutral morphologies. So far as one can judge, each is equally effective at dispersing spores. There are modest food niches and possible competition among the separate amoebae of different species, but it is difficult to argue that this might translate to competition between them in their multicellular stages.

Micro Living Fossils

One of the exciting recent developments in cellular slime mold biology (coming from the laboratories of S. Baldauf and P. Schaap and their collaborators) is the construction of molecular phylogenetic trees.[22] Based on one or two genes, many of the existing species were analyzed and it was possible to show that

[22] Schaap et al. 2006; Romeralo et al. 2011.

they fall into four major groups, and those groups could be placed in a sequence: group 4 is the most recent, and group 1 the most ancient. With the exception of differences in branching patterns, their morphologies do not show great distinctions, but slight variations around the basic body plan of a mass of spores held up into the air by one or more slender stalks.

In contrast to this morphological sameness (as compared to the variety of shapes in radiolaria or diatoms), biochemical differences between species appear to be more dramatic, although our knowledge of this diversity remains rather rudimentary. One fascinating glimpse comes from the work of Pauline Schaap and her colleagues.[23] They compared the proteins known to be activated by cyclic AMP in an ancient and a modern species. The latter had four such proteins; the ancient species only one.[24] Furthermore, the ancient one was not for the aggregation by cyclic AMP, but for a cyclic AMP that is involved in a signaling step that occurs later in development. The primitive species, *Dictyostelium minutum*, uses folic acid for its aggregation attractant, a substance also given off by bacteria, and attracts the feeding amoebae. In other words, the chemical system that attracted the amoebae to food seems to have been co-opted for aggregation in descendant species. There has been an evolution in the chemical signaling that is not reflected in the morphology.

[23] Alvarez-Curto 2005.
[24] Parent and Devreotes 1996.

The main difference between *D. minutum* and the more modern *D. mucoroides* is that the latter is larger than the former. Both have a single stalk and a terminal mass of spores. The biochemical changes have become altered over time, barely affecting shape. It is a different manifestation of de Beer's concept of "clandestine evolution."[25] He thought of the adult insects of virtually identical morphology whose larval stages showed radical morphological differences. Here I am talking about biochemical differences, or changes, that are not affecting the morphology of any stage of development.

What is different in these slime mold phylogenies, and those of other microorganisms, from those of big animals and plants is that in the latter we see great evidence of progressive extinctions in their fossil records. This is true of the old familiar phylogenies of horses and dinosaurs. In contrast, many of the ancestors in the slime mold phylogenetic tree are still alive today! One might argue that we have no fossil record of them, but by analogy, as we have seen, radiolaria and diatoms have left an extensive fossil record, and often there has been little or no change in their morphology over many millions of years. No doubt they also have a biochemical clandestine evolution, but that has yet to be examined.

---

[25] De Beer 1962. A similar example has been revealed by T. Winkler and his group (Asghar et al. 2012). They find that the dipeptide glorin, which is known to be the aggregation chemoattractant for *Polysphondylium*, also exists in other, more ancient species where it plays a different role and activates developmental genes.

There are interesting exceptions among large organisms that were dubbed "living fossils" by Darwin. The morphology of horseshoe crabs has remained much the same for at least some 250 million years, as with the recently discovered fish *Latmeria*, whose coelocanth relatives became extinct roughly 400 million years ago. The same story is found in plants: the ginkgo and metasequoia trees' main relatives also went extinct eons ago. But these examples are rare.

In the cellular slime mold phylogenetic tree, however, many of the early forms continue to exist today. Clearly they too are living fossils, and they are not the exceptions but the norm. One possible reason for their continued existence might be that their morphology is little affected by natural selection. And my hypothesis is that this is so because they are small.

But there is a curious twist to this story. Their morphology may have changed little, but their biochemistry has. One would assume that natural selection is responsible for the general shape of the fruiting bodies, which is so effective in the vital process of dispersal, and that the biochemical evolution is also the result of selection for improved efficiency. But why have not the ancestral, and presumably less efficient species gone extinct? We can only guess at the answer to this question, and here are some possibilities.

It might be that the advantage of a new biochemistry is so modest that the old is not outcompeted. Or it might be that through the highly effective dispersal the new inventions become physically isolated from the old, and in their isolation individuals with the old biochemical networks continue as they always have. This fits in with the idea that microorganisms have enormous capabilities for dispersal, and different strains will constantly be separated from one another. Being small is the key to there being so many extant living fossils among the cellular slime molds. This is consistent with the hypothesis that the degree of natural selection is greatly attenuated for these small organisms, and as a result the advantage, the improvement of the new biochemical variant, is very modest. Or the advantages of the new cannot lead to the extinction of the old because they become so isolated physically through their highly effective dispersal mechanisms. The different genetic variants get separated from one another because they are spread over such distant and isolated areas in the surrounding soil, or even farther if they are carried in the gut of animals, and even of migrating birds over thousands of miles.[26] Either explanation—or both—could account for micro living fossils.

The novel genetic changes, along with the ancestral genetic constitution, have a better chance of surviving because there is a high probability that they might fall on separate patches of bacteria to start the next generation. This is not the classical drift of genes, but whole genomes. It is an ecological isolating mechanism that would favor the maintenance

[26] For a brief review and references, see Bonner 2009, pp. 38–42.

of genetically divergent cells and the fruiting bodies into which they develop.

Fungi: Molds in the Soil

There is also the interesting possibility that small fungi, or molds, that inhabit the soil might have nearly neutral morphologies. As in cellular slime molds, their fruiting body is their prime agent for spore dispersal.

Dispersal of this kind is a widespread phenomenon, and of course has arisen by natural selection. If one examines the surface of soil in a moist chamber, one cannot but be impressed by the number of small fruiting bodies that appear on that surface. Almost invariably they will be some species of fungus, for there are a vast number of species of mold that have filaments that rise to hold small terminal masses of spores. For instance, one finds common species of phycomycetes, such as *Mucor*, or *Rhizopus*, or species of the innumerable *fungi imperfecti*, such as *Aspergillus* or *Penicillium*, and a vast number of others. It is perhaps the most massive collection of examples of convergent evolution among all living organisms; the minute fruiting body has been reinvented over and over again in totally disparate groups of soil organisms—not only in slime molds and filamentous fungi, but also in bacteria (the stalked, multicellular myxobacteria), and there is even one example among ciliate protozoa. One can only conclude that the selection pressure for spore dispersal among small soil organisms must be intense, and the same solution has been independently reinvented many times.

There is evidence that soil contains an impressive number of fungal molds that exist side by side. For instance, Nanjundiah and his colleagues have examined a forest soil in India, where, during the wet seasons, they estimated that within an area of 16 hectares (400 square meters) there are more than two hundred species (members of roughly fifty genera).[27] Again it is difficult to imagine that this plenitude does not involve a considerable degree of neutrality in their morphologies.

The neutral morphology hypothesis is a way of accounting for a number of features found among small organisms, in particular the incredible diversity in protists found within a single environment. This diversity among microbial phenotypes is a phenomenon at the fringe of Darwinian evolutionary biology that might be better understood from an unconventional perspective. Neutral morphologies are a possibility that cannot be ignored—it may provide valuable insights into the world of micro eukaryotes and their particular evolutionary processes.

[27] Satish et al. 2007.

CHAPTER 4

The Evolution of the Decrease of Randomness

An important way of reducing the effect of randomness is to become bigger, as have all large multicellular organisms. The results of random mutations are filtered by the vast number of steps they go through to produce an adult animal or plant. Each step goes under the absolute scrutiny of "internal selection"; there are virtually no opportunities for any deleterious change to survive this sequence of steps that we know as development.

This does not mean that every step, every final morphology is determined genetically, nor is it determined by chance. As D'Arcy Thompson[28] and some current authors[29] have pointed out, often physical forces, for instance surface tension and adhesion, cap all other influences and can play a major role in the formation of particular morphologies.

---

[28] *On Growth and Form* (1917 et seq.)
[29] For example, Newman et al. 2006.

Large organisms are unlikely to have an overall neutral morphology like small ones, and the reason is to be found in their elaborate development. The greater the size, the more developmental steps. The voyage from a single cell, a fertilized egg, to a large, mature organism with millions of cells is a process that cannot be chaotic, but must be controlled if it is to achieve a consistent ultimate shape from generation to generation. There can be no significant deviation from those set steps to get from one generation to the next.

Let us look at those multitudinous steps in some detail.

The Origins of Multicellularity

What we want to do is examine in some detail the creation of complexity—to follow the steps involved. A good place to do this is in the origin of multicellularity. Any multicellular organism we look at today is so complex we cannot help but wonder how it got that way. We do not accept the idea that it arose full blown from the head of Zeus; it must have had a beginning. Furthermore, as I have already pointed out, there were numerous origins, for multicellularity was invented a number of times. The great difficulty is that all those well-established origins occurred eons ago, and the only way we can reconstruct them is by hypothesis, by bald guesswork. (Of course there could be new inventions of multicellularity occurring today, but how could we ever find them?) Yet such thought experiments can give us some general idea of how complexity might have evolved and therefore some understanding of how it continues to do so.

The current approach to the complexities of multicellularity is largely to find a way of reducing that complexity to simple rules. This is "systems analysis," and mathematical modeling is one of the main tools to make sense of the great tangle of information. The very same approach has been used to simplify the complexity of the brain with continuing effect. This is indeed the wave of the future and should be vigorously encouraged. It is a top-down approach trying to sort out all the morass of details so that we can understand the composition of multicellularity and how it ticks. There are numerous enterprises all over the globe that are pursuing this approach and new light has been, and will continue to be, shed. The future is full of hope.

Here I would like to take a much more modest approach that goes in the opposite direction: it is a bottom-up tack.[30] The logic behind this approach is that the first steps in any advance in complexity, such as the invention of multicellularity, must have been direct and simple—the complications came latter. It asks what are the minimum events needed to achieve a step forward in complexity, and such an approach, even if it is highly speculative, will provide some insight into the bare essentials necessary

[30] For a fuller discussion, see Bonner 2000.

for such a step. Instead of seeking the fundamental basis of a step in the development of complexity by analyzing the system as it exists today, I take a minimalist approach and simply ask, how might it have begun? This is another way of exposing the bare bones. Both approaches are imperfect: the bottom-up way rests on the perilous ground of speculation, but so does the top-down way, for any analysis trying to organize vast quantities of details involves uncertainties. The mathematical models are riddled with assumptions and guesses. Yet both the top-down and the bottom-up modes of illuminating the evolution of complexity will help us to understand the basic principles of how it takes place.

As already mentioned, in early evolution there must have been a persistent selection for size increase, which was achieved independently a number of times by becoming multicellular. It was an easy and obvious way to become bigger.

The First Step towards Multicellularity

The first step is undoubtedly for the cells to become sticky so that they can adhere to one another. One of the most dramatic demonstration of this comes from an elegant experiment by E. M. Boraas and his colleagues.[31] It is often argued that if an aquatic form becomes larger, then it is protected because it is too big to eat, and this might be a reason for the natural

[31] Borass et al. 1998.

selection of size increase. Alongside is the reciprocal argument that the larger the beast, the more effective it is as a predator. I had always assumed that this had never been tested by experiment, but I was wrong. A few years ago Boraas and his colleagues grew two organisms in the laboratory in continuous culture. One was a predator (a carnivorous unicellular flagellate) that was added to a pure culture of a small, unicellular alga. They remained in equilibrium, that is, in the same proportion; the rate of being eaten and eating remained constant. This was shattered when suddenly the larger predator lost its foothold and slowly disappeared, for the small prey could no longer be conquered. What had happened was that the small alga mutated and produced a sticky extracellular substance causing the cells to clump, and the clumps were too large to be engulfed. It is presumably a single mutation that allowed the formerly susceptible prey to avoid being eaten. This is a beautiful demonstration of the advantage of becoming multicellular and the role of size in the prey-predator struggle.

Obviously, a small clump of algal cell is not the same as an organized multicellular organism. Being photosynthetic for energy, and provided they did not get too large, they could persist, although we can safely assume that their existence would be perilous and improvements for survival would again be favored by selection. The stickiness and the resulting clumping must be considered an unstable condition that will be further advanced by more mutations and gene arrangements that consolidate the initial gain.

The Steps That Follow

Without worrying about the sequence of which came first, let's consider some improvements that could have occurred by mutation and gene rearrangements that would be adaptive and therefore retained. A special group of cells at the outside edge of the clump might arise to keep the cells in a coherent unit. This might be coupled with some system to control the cell number, and therefore the size of the clump. One possible way to control this is for mutations that limit the number of cell divisions. Some innovations might be associated with a shape that is optimal for gathering the Sun's rays for photosynthesis. And then some new tricks might appear to facilitate either or both asexual and sexual reproduction. This is a minimum list only to illustrate the kinds of changes that must have occurred. To see what an enormous number of innovations have in fact occurred, one need only look at the rich array of different kinds of small, multicellular algae. They are successful and exist in abundance in a variety of forms today.

The Cellular Slime Molds

What I would like to do now is examine another case of independently invented multicellularity in more detail. My choice is the cellular slime molds, largely because I have studied them for many years and have a clearer idea of what possibly might have been their early steps of increasing complexity. They have the further advantage that even in their most highly developed form they are very simple organisms. For instance, at the pinnacle of their complexity they only have two cell types. Before I begin it might be helpful to remind the reader of their peculiar life cycle (fig. 7).

Cellular slime molds are soil amoebae that feed on bacteria. Once they have cleaned an area of food, and after a few hours of deprivation, they come together to form a clump of amoebae. One or a few of the amoebae start to secrete a chemical attractant, and all the outlying amoebae respond to it by streaming towards the attractant-secreting cells, and in doing so begin to secrete the attractant themselves—it is a relay. In the species most studied, *Dictyostelium discoideum*, which for convenience I will describe here, the collected group of cells forms a bullet-shaped slug with a front end and a hind end, and it is encased in a slime sheath. The slug moves and migrates to the surface of the soil, a location that is optimal for dispersal. Once there, it forms a fruiting body: the anterior amoebae of the slug transform into stalk cells producing a stalk that rises into the air, lifting up the posterior amoebae of the slug that become resistant spores (fig. 7).

We begin, of course, with extremely complex eukaryotic cells: the amoebae. Many dangers for them exist in the soil, but perhaps the most universal are environmental: desiccation in times of drought and freezing in wintertime. Soil amoebae in general have

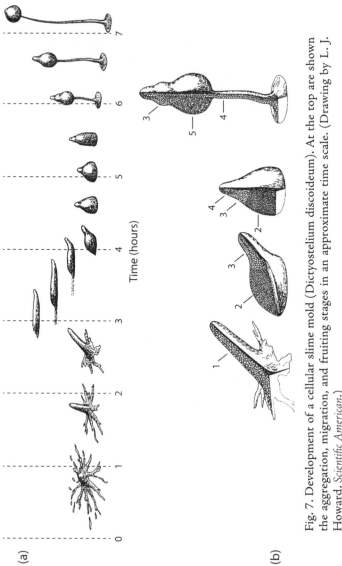

Fig. 7. Development of a cellular slime mold (Dictyostelium discoideum). At the top are shown the aggregation, migration, and fruiting stages in an approximate time scale. (Drawing by L. J. Howard, *Scientific American*.)

(a)

Time (hours)

(b)

THE DECREASE OF RANDOMNESS   71

a resistant stage where a single amoeba becomes encapsulated into a dormant, resistant cyst. Even many species of cellular slime molds are capable of bypassing aggregation and fruiting, and individual amoebae will encyst if the environmental conditions become adverse. Therefore, forming resistant bodies is a process that probably had already been present before the arrival of multicellularity.

There is a curious group of amoebae, the protostelids, whose solitary amoebae not only encyst but secrete a minute stalk as they do so that raises the single spore up into the air (fig. 8). From the point of view of natural selection there is an easy reason why such a mini stalk might be adaptive. For the reproductive success of soil amoebae, dispersal is clearly key. Bacterial food sources are scattered in patches in the soil, and survival of the amoebae depends upon being able to find those patches. We know from direct observation that the common way of achieving this task is through the transfer of spores from one area to another by one of the many invertebrates that crawl over the surface of the soil: insects of various sorts, spiders, mites, small millipedes, worms, and so forth.

The chances of a spore sticking to a passing animal will be greatly increased if it is raised up into the air on a stalk. It follows that any mechanism that increases (within limits) the height of the stalk would increase the chances of catching a ride. The one way this might most easily be achieved is by becoming larger through multicellularity.

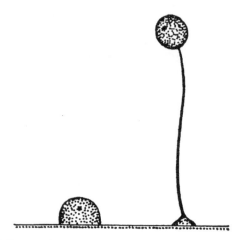

Fig. 8. Protostelium. A single amoeba secretes a thin stalk and rises into the air. (From Grell 1973; drawing by G. Gerisch)

Unfortunately we have no idea how this might have arisen in soil amoebae. For the very beginning, there are two possibilities. One is that like Boraas's algae they became sticky after feeding and form a clump in the soil. The other is that the amoebae attract one another by chemotaxis. As mentioned earlier, it is known that amoebae are attracted to bacterial food; the bacteria give off substances, such as folic acid, and the amoebae will move towards them by this chemical attraction. As single cells they have a built-in mechanism to orient in chemical gradients, and to modify this ancestral mechanism to attract one another might not involve too great a modification of the genetic background. An additional form of chemotaxis could easily be incorporated. In many of the species of slime mold, the attracting chemical is cyclic AMP (cyclic adenosine mono-phosphate), a small molecule found in the cells of our own bodies.

There is an added feature that is necessary: having the cells attract one another would produce clumps, but that is not enough. There must be some organization; one wants to have leaders and followers. This might be a relatively easy step because if one, or a few, of the amoebae started secreting the attractant sooner than the others, it would lead to an antero-posterior organization to the cell mass. Added to this is the relay system previously mentioned, where the cyclic AMP attractant induces other cells to produce it and causes a wave of attraction that passes down a chain of amoebae. In this way the reach of attracting amoebae can be increased. From an evolutionary point of view, at least two innovations are required. One is a new attracting chemical that does not just mean food, and the other is a relay system where the new chemical stimulates its own production in neighboring amoebae, that is, it is autocatalytic. Therefore, we assume that in the beginning there was selection for the mutual attraction between amoebae, and that this chemical had to be autocatalytic.

Before we get to the genetic basis of these innovations, let me continue this hypothetical descriptive story until we reach an organism that looks something like a cellular slime mold that lives today. The amoebae aggregate by chemotaxis and, in the clump that forms, the motile amoebae must adhere to one another. This is achieved in two ways. One is that they produce suitable molecules at their surface that make them sticky so they adhere to one another, a step quite similar to what Boraas found in his mutant alga. The other is to produce a slime sheath that encloses all the sticky cells, a further step in the organization of the multicellular mass. It is quite possible the two phenomena are part of the same mechanism. The antero-posterior polarity of this motile slug is strictly retained, undoubtedly because of the properties of the chemotactic system.

Two major steps remain to be taken to produce a modern cellular slime mold. These steps are interconnected. First, a structure that will raise the mass of spores (sorus) up into the air is essential and involves a moving of the cells to form a fruiting body. The dynamics of this process are complex and therefore it no doubt consists of numerous separately controlled substeps. As a result, it is difficult to dream up an obvious hypothetical evolutionary—selection driven—sequence. The other major innovation taking place is going from one cell type to two. Again, this must have involved numerous substeps— equally difficult to imagine how they evolved, and how they are intertwined with the complex morphogenesis of the fruiting body.

There are a few checks and balances to these events that are important. For instance, the attraction-relay system could not operate successfully without a countercheck. It is known that there is an inhibitor enzyme that controls aggregation chemotaxis by breaking down the attractant, and for obvious reasons this is necessary. If it were not broken down it would become too concentrated and no longer able to produce an effective gradient to guide the amoebae. There is even a further regulatory substance that controls the enzyme by inhibiting it. We have therefore in this one system—the aggregation of amoebae into clumps—three key regulators: the attractant, the attractant remover, and a repressor of this remover. All this fine-tunes the aggregation process—to make it more reliable and consistent.

We assume that from an evolutionary point of view these events did not all arise at the same time, and the sequence might have been something like this:

Chemotaxis for food.
Chemotaxis between amoebae leading to the
    recruitment of a new attractant molecule.
An enzyme that breaks down the aggregation
    attractant so that it can be more effective at
    optimal concentrations.

An inhibitor of this inhibitor evolved as a safety valve so that the final concentration is optimally appropriate.

The beginning step for multicellularity for cellular slime molds was the dual role of cell adhesion and the far more important chemical attraction between the cells. This was followed by all the checks and balances just described. Therefore, the invention of multicellularity became more controlled and reliable. No doubt this occurred because each of the successive innovative steps in this growing complexity was either redundant, and therefore persevered by being under the same selection of the original innovation, or it increased the efficiency, and therefore had the selective advantage of the original step.

Let us now examine the second event in slime mold multicellularity: the formation of the two cell types, stalk cells and spores. Circumventing the details of how this was discovered, in *Dictyostelium discoideum* it has long been known that the anterior amoebae in the migrating slug will become stalk cells, and spores come from the posterior amoebae.[32] They differ in one important respect: how well fed they are. While feeding on bacteria, the amoebae divide by binary fission every few hours; therefore, at the moment of aggregation they differ in how much food they have inside them. During aggregation the amoebae sort out, and those with little food move more rapidly to the front end to become stalk cells, while the well-fed ones lag behind and become spores. (It is of interest, but perhaps not completely relevant here, that by experiment one can always convert prestalk amoebae into prespore ones, and vice versa. Their fates are not fixed until their final maturity.) The two cell types could not be more different; the stalk cells swell with big vacuoles and die, while the prespore amoebae each become encapsulated, and upon germination they are capable of starting a new gen-

---

[32] For a review and references, see Bonner 2009.

eration. The first clue to their ultimate fate depends upon on how well fed they are, so from this slight difference they produce two very distinct cell types. (There is an interesting parallel here with social insects, where differences in nutrition of the developing workers lead to different size castes who differentiate by dividing the labor. We will return to this point later.)

The Development of Larger Forms

Let us now examine the development of more conventional animals and plants. They are usually the result of sexual reproduction, which is almost universal in bigger forms. It involves fertilization and the formation of a single-cell zygote. That single cell then undergoes a vast number of divisions to produce a mammal, or a fish, or a tree. Let us pause for a moment to consider—in the most general terms— how this is achieved.

There are a few basic processes that are part of the development of all organisms. They all begin with the establishment of a polarity—a direction, a front and a back—followed closely by the formation of a gradient. These are the first steps in organizing the embryo. They are followed by the formation of different regions within the embryo that begin to show distinctive characteristics, be it a limb of a frog, or a leaf of a maple tree. The changes are accompanied by continuing cell divisions, and as they grow there will arise differences in cell shape and cell biochemistry, part of an intricate and elaborate division of labor.

Such a description is so brief, so superficial that it fails to show the complexity, and the diversity, of the different kinds of development. One need only look at a modern textbook of developmental biology to begin to appreciate the magnitude of the process. It is a phenomenally gigantic bit of construction, and it varies, sometimes quite radically, for different organisms. It involves an incredible number of steps, one following the other with uncanny accuracy. There can be minor missteps, but they are rare: major ones would lead to the death of the embryo. For every plant and every animal that exists on Earth today, the entire, great sequence of events has to be consistent and approach perfection, otherwise it would be stopped along the way and adulthood would never be reached. This is the "internal selection" mentioned previously.

This rigidity in the control of the sequence of development will make a random change in the form of a mutation anywhere in the sequence extremely unlikely to survive, unless it occurred at the terminal end of the sequence.

The Genes That Control Multicellularity

This brings us directly to the genetic foundation of the steps. Each additional change towards multicellularity was ultimately the result of a mutation: not any mutation, for indeed mutations are random, but those that were retained by natural selection because of the advantages that might accompany an increase in size.

Initially there must have been a single mutation to trigger the first step. Then came the continuous arrival of new mutations, and as already mentioned, some repeat or duplicate the original one, and others refine it so that it becomes more consistent and more resistant to being disrupted by perturbations. If natural selection significantly favors size increase, then all these kinds of mutations that are involved in the creation of multicellularity will be favored and retained. Obviously it is not just the mutations themselves that are key, but what follows in the way of the interaction of the genes, both new and old, and the consequences of their activity. As Eric Davidson has shown, a significant facet of these interactions is the invention of regulatory genes that control the activities of the other genes involved in development.[33]

An important point should be made here. Even though we have started with the first steps towards multicellularity so that we can see its formation in the simplest terms, clearly the process became remarkably complex after the beginning. To understand this we need to add the perspective of time. Cellular slime molds are very ancient, and since their beginning they have gone through a staggering number of generations. Based on the arrival

[33] Davidson 2009.

of soil on the surface of the Earth, and based on the presence of small invertebrates in that soil, the first cellular slime molds probably arose somewhere in the middle of the Ordovician, at least 400 million years ago. Since those crawling invertebrates are the prime agents of spore dispersal, we can assume a strong selection pressure for fruiting bodies that stick up into the air that can plant sticky spores on the sides or the limbs of beasts as they pass. With the continuing arrival of new mutations there is enough time to produce a network of genes that all work towards the same phenotypic objective. Endless time has been the key to foster ever-increasing complexity.

Gene Nets

With the luxury of that endless time, and the steady rate of the appearance of new genes by mutation, the number of genes directly involved in determining a particular step will continue to increase. As already pointed out, some of those mutations will not do exactly the same thing as the original mutation, but will affect it, perhaps by modifying its effect or controlling it in some way. This means that the genes and their products affect one another and inevitably fall into what can be called a network. Because of the network, the action of any particular gene can affect the action of one or more others, the whole mechanism of genetic control rises to a new level of complexity. To see how this might be, in the example just given of slime mold chemotaxis we saw that on the gene product level there is a regulating system that is clearly evident: the initial attraction, the auto-catalytic relay that gives the attraction direction, the enzyme that destroys the attractant so that it maintains an effective concentration gradient, and an inhibitor of that enzyme that keeps it from getting too enthusiastic and wiping out the primary attractant completely. In other words, the control of the key process of aggregation has evolved and a gene net is the result.

This is only one example for the slime mold. There are no doubt similar interacting gene networks that control not only the migration movement in its many aspects, but also the movement involved in lifting the fruiting body up into the air and all the details of how some of the amoebae become stalk cells and others spores. Perhaps the surprising thing about this complexity is that it exists in such a simple organism, but remember that it has had perhaps 400 million years to invent and fine-tune its gene nets.

That complexity increases with time is illustrated in some elegant discoveries of the evolution of chemotaxis in the cellular slime molds that were mentioned previously. To begin, we now have a molecular phylogenetic tree that gives us the ancestry of different species.[34] We also know that in the modern species *D. discoideum* there are four known receptors that can specifically combine with cyclic AMP.

[34] Schaap et al. 2006, Romeralo et al. 2011.

In each case this leads to the activation of a chemical pathway that leads to a different result.[35] The first that appears is the one already described—the one that leads to the relay by causing the internal production and secretion of cyclic AMP by autocatalysis. The other three receptors stimulate internal chemical chains that in turn stimulate other specific pathways involved in cell differentiation and other aspects of development.

Not all these receptors are present in the ancestral species *D. minutum*; in fact, it possesses only one that is involved in the process of differentiation.[36] From this we can infer that originally cyclic AMP was not the attractant but was co-opted for that role at some later time in the species' evolution. This is an example of clandestine evolution referred to earlier, where a biochemical evolution took place that has not been reflected in the morphological evolution.

It is of particular interest that the aggregation attractant of *D. minutum* is folic acid, which, as we saw, is one of the substances given off by bacterial food and leads the amoebae to their prey. This reinforces the idea put forward earlier that the aggregation chemotaxis is derived from the feeding chemotaxis. This example shows that, with sufficient time, there is an evolution of the biochemical pathways that are involved in the evolution of multicellularity, which means there has been an evolution of gene

[35] Review: Bonner 2009.
[36] Alvarez-Curto et al. 2005.

nets. Each event just described is part of a gene net that keeps changing over great spans of time.

Modules

There is evidence, even in slime molds, that these nets can to some degree coalesce and become more integrated. These unified gene nets might be called *modules*, to dignify them as the next step in the evolution of multicellularity. That they are at a different level of cohesiveness is illustrated in the phenomenon of heterochrony, well known in higher animals and plants. Let me give a classical example.

Most salamanders, before living on land, pass through a larval aquatic stage in their life history. In the transition they go through metamorphosis, which involves losing their gills and developing lungs. They also come to sexual maturity on land, mate, and produce fertile eggs that are returned to the water. There is an interesting exception: *Axolotl*, a large Mexican salamander, becomes sexually mature while still in the water, even though it has gills. In other words, the gene-net-complex responsible for sexual activity can be independent of the one responsible for the gill-lung transition; they can act at different times with respect to one another. The temporal aspect of this independence of the two phenomena is appropriately called heterochrony, because it is their timing—the time when they occur—which varies. Their gene nets are gathered into modules that to some extent can go their separate ways.

Heterochrony can also be seen in the cellular slime molds. There is a striking difference in the progression of differentiation between *D. discoideum* and its relatives and the species of *Polysphondylium*. The fruiting bodies of the two genera differ in that *Dictyostelium* has a single stalk with a globular sorus at its tip, while *Polysphondylium* has elegant whorls of mini stalks along its axis (fig. 6). They arise by leaving behind successive deposits of rings of cells on the stalk as the fruiting body rises into the air; the whorls of tiny fruiting bodies sprout from those rings. In *Dictyostelium* the partial differentiation of the prestalk and prespore amoebae arises during migration, but obviously this would present difficulties for *Polysphondylium* since the final spore differentiation occurs separately in each of the little fruiting bodies in the whorls. Therefore, the prestalk-prespore beginnings of differentiation must be delayed so that each of the small secondary fruiting bodies can independently develop stalk cells and spores. This is precisely what happens: the amoebae of *Polysphondylium* all along the rising stalk show no signs of early differentiation as they do in *Dictyostelium*, but at the last moment they rush through the differentiation process into their mature stalk cell and spore mode. It is clear that the steps that lead to this final differentiation are essentially the same, but they differ in their timing.[37] In *Dictyostelium* it

[37] Review: Bonner 2009.

is spread out over time; in *Polysphondylium* it happens all at once at the very last moment. This a good example of heterochrony, for the modules that lead to stalk cell and spore differentiation are dissociable and can have different timings. This heterochrony has made it possible to have two quite different morphologies even though the basic mechanisms of differentiation are the same.

In this instance we can say that the gene nets have become integrated into modules because they exhibit the ability to act independently to some degree. There is a tighter integration between the nets that allows them to be called modules. It must be remembered that modules and nets coexist, and even solitary genes that have no connections with others can be present as well. In fact, these early forays into multicellularity soon become a great jungle of genes and their products. So, very early in the evolution of multicellularity, with the help of great spans of time, the complexity accumulates to a remarkable degree.

Becoming Bigger

Consider briefly what happens in larger animals and plants. The appearance of new mutations, new gene nets, and new modules continues; there is a vast and steady increase in their number, resulting in an overall increase in complexity. If it is difficult to unscramble the molecular details of the relatively simple slime molds, imagine the problem with a large organism. Think of its extended development that may continue for long periods of time, and each stage of that development will have a succession of many new genes, new nets, and new modules. I said for the small slime molds that these components were sufficiently complicated to make a jungle; with large forms we have a jungle of jungles. The complexity becomes beyond our imagination.

It is always hoped that it will be possible to find some simple basic rules or phenomena that will illuminate the darkness, and indeed this has occasionally happened. For instance, we have the HOX genes that are omnipresent among animals and in different ways are responsible for laying down the principal parts of the organism. There are also the PAX genes that are at the basis of the production of eyes throughout the animal kingdom, no matter how different the eyes are in their construction. Furthermore, there have been important advances in the analysis of regulatory genes and the role they play in development, as previously mentioned. Yet, while this has been great progress, it only chips away at the edges of the tangle of the great complexity.

Size and Natural Selection

It is not my intention to pursue these big problems here. I only raised them as background to the main question I want to address: How are these different levels of complexity that are found in different-size organisms—from the smallest and simplest to the largest and most complex—affected by natural selection? In evolutionary thinking we tend to treat them the same way, but I want to show that size might make a big difference.

I will illustrate why small and large organisms are affected differently by describing a simple diagram, and providing a relevant analogy.

The Diagram

In fig. 8 the bifurcating lines are meant to indicate gene nets, and they are connected to one another. This is not a realistic depiction, but merely a shorthand way of representing the advancing steps of development. These steps are interconnected and sequential, building one upon another. In a small organism, with a minimum number of steps, a mutation will almost immediately be at the end of the chain and therefore able to affect the morphology of the whole organism. This condition was examined in some detail in chapter 3. In a larger organism the interconnected lines that represent the incremental steps now become more numerous, corresponding with the degree of the increase in size. Such a simple diagram shows none of the complexity that goes with size; for instance, in a mammal the formation of limbs, or the formation of organs such as the heart, or the kidneys, or even the appearance of modules. All of those key advances are subsumed among the great mass of connected lines.

Now here is the key point. A mutation could arise at any point in the thicket of connected lines, but if a step were altered through a debilitating mutation, the sequence of steps that would have followed to complete development would be blocked, and the result would be the death of the embryo. They would be eliminated by the internal selection of Lancelot Law Whyte, as discussed earlier. Only the mutations that appear at the end of the chain would have a chance of survival. They would become part of a grown, mature organism and will now come under the scrutiny of natural selection to see if it will be fruitful and give rise to offspring (fig. 9).

Natural selection depends on variations in the phenotype determined by the genes, and those phenotypes that are most successful in producing offspring will persist, as will, of course, those genes. This is standard neo-Darwinism—the marrying of natural selection with genetics. It means that any mutation that leads to better vision, or greater running speed, or superior camouflage, and so forth will lead to reproductive success and will be retained in the genome. In the case of a large, complex organism, mutation can occur in any part of the mature organism. If the change affects that whole organism favorably, it will be retained; if it has adverse effects, it will be eliminated. It means there are a vast number of ways happenstance mutation can tinker, and some of those changes may benefit the whole organism.

To give an example, of among many, we can turn to Darwin's finches in the Galapagos. The diet of the seed-eating species may vary depending upon

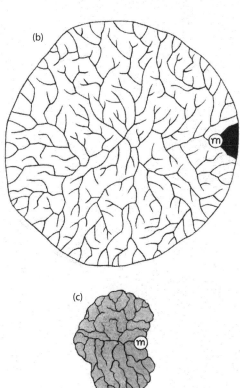

Fig. 9. An over-simplified diagram showing how a mutation in a microorganism (a) can affect the entire morphology of the resulting phenotype, while in a larger organism (b) a mutation may only modify a small portion of the phenotype. If a deleterious mutation arises in the middle of development, the embryo dies (c). The branching lines represent the sequential changes during development, such as the progressive steps in the determination of cell fates. (Drawing by Hannah Bonner.)

the environmental conditions, for the seeds may be abundant and small after a rainy season, but after a prolonged drought only larger seeds are available. Birds with small beaks are most effective in consuming small seeds, and large seeds can only be cracked open by large beaks. The selection for beak size will therefore vary in dry and wet climate periods, and these trends towards large and small seeds, with the corresponding changes in beak size, can be followed through many seasons, as Peter and Rosemary Grant have done in their exemplary studies; they have caught natural selection in the act.[38] Furthermore, some of the genes connected with beak size and shape have been identified.

An Analogy

In order to visualize how viable random events (mutations) appear mainly at the tail end of development, I want to compare the development of organisms to boat building. The big differences are obvious: a boat is not a living organism and it does not have mutations. For the equivalent substitute to "living," invoke the men who do the building—the carpenters, metal workers, the plumbers, the electricians, and so forth. For the mutations, substitute construction errors, failure of materials (for example rotten wood), and inadvertent deviation from the initial plan.

[38] P. and R. Grant 2008.

Consider the two extremes: a small rowboat and a huge battleship. Perhaps the most obvious difference between the two involves time: a rowboat can be built in a day; a battleship will take a year or more. One finds the same difference between a single-cell protozoan and an elephant: a matter of a few hours for the protozoan, and ten years for an elephant to reach maturity. The rowboat requires putting a few planks together to make the right shape, while a battleship requires welding together huge steel plates, thick enough and strong enough to hold the engine, all the rooms and compartments, and the heavy guns. In both, as in living organisms, they need a skeleton to hold them up and together. The rowboat has a wooden exoskeleton; and in a battleship it is made of steel, but it also has a steel endoskeleton to support its internal parts. The protozoan equivalent is a cell wall that gives it shape, and the battleship equivalent is lobster shells and the bones of elephants and huge dinosaurs.

And now for the mutations. An error in the rowboat building might be a rotten plank that leads to the rowboat filling up with water. In the battleship it might be any error in the construction of the steel shell, in the internal struts, in the plumbing for the kitchen or bathrooms, in the vast electrical system for all the rooms, and to the navigating instruments. Should a malfunction of some key part occur, the big ship could be made nonfunctional.

The big difference between boats and living organisms is that when any part needs repair during construction, workmen will swarm to the troubled spot and fix it. A living organism has no such recourse. If a mutation adversely affects a step during the course of development, everything stops and the embryo dies. This is the internal selection that culls as development proceeds.

One more point about the analogy. In an organism any mutation that occurs during the last stages of development may survive in the final beast, and it will remain in subsequent generations if it is favored by natural selection. This is obviously also true for the battleship. If in the final painting of the ship everything is in the proper gray, except for a spot of bright red, the spot cannot remain. It is not removed by natural selection, but by the irate order of the admiral. And it is much easier to right such a superficial error than something more fundamental in the bowels of the ship.

If one compares a small organism with a large one, random mutations are immediately evident in the morphology of the adult. In larger forms the mutations that occur during development are not only eliminated if they have negative effects, but they are most likely to cause the demise of the whole embryo, that is, by internal selection. The consequence is that in larger forms less randomness survives than in microorganisms.

CHAPTER 5

An Exception

Where Small Organisms Suppress
Randomness

The point has already been made that the sexual system has been burnished by natural selection so that in each generation the degree of variation in the offspring is optimal—not too little and not too much—which allows natural selection to take place and makes evolutionary progress possible. An important size-related phenomenon is that in many simpler, small organisms this sexual variation-control mechanism can be turned on and off: periods of sexual reproduction will be interspersed with periods of asexual reproduction. In larger forms (with some rare exceptions) only the sexual route is possible. And it is only in the sexual cycle through recombination that the all-important controlled variation that is so essential for evolutionary progress is produced. The asexual reproduction found in many small organisms, which can usually be very rapid, is available to exploit particularly favorable conditions as quickly as possible. It rushes in when the environment is just right and allows fast multiplication before unfavorable conditions close it down.

This principle appears to be at odds with the contention that randomness is more common in small organisms, for asexual cycles shut off randomness; it is only found in the sexual ones. The ability to switch to asexual cycles is something denied to larger forms. It is an adaptation to leap forward and take advantage of the arrival of favorable external conditions.

In some organisms there is another reason for the alternation of sexual and asexual cycles. An extended series of asexual generations may result in some sort of requirement for rejuvenation, and this is achieved by sexual fusion. For example, there is the interesting case of diatoms where each asexual division involves the separation of the two halves of their silica box and the synthesis of a new bottom for both daughter cells. Since it is always the smaller bottom that is newly formed at each cell division, the average size in the population becomes smaller and smaller (fig. 10). Eventually a threshold will be reached and sexual reproduction intervenes. All four of the old shells are shed as the contents of two cells fuse, and the large zygote builds a much bigger and entirely new lid and bottom.

The important point is that this ability to switch from sexual to asexual cycles is found primarily in

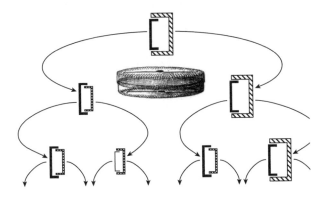

Fig. 10. A diagram showing the progressive decrease in the size of a diatom. The central drawing shows how the valves fit together, and since after cell division each valve develops a new "bottom" which fits inside, the mean size of the population decreases. The size diminution in the diagram has been greatly exaggerated to illustrate the point. (From Bonner 1965; drawing by Patricia Collins)

small organisms; almost all higher forms have only the sexual mode of reproduction. This is another instance where size is a key factor. As we shall see, there are obvious reasons why this is so; it is another example where size dictates what is and is not possible, as it did for the relation between size increase and the obligatory increase in complexity.

The alternation of sexual and asexual cycles in small organisms is often made possible by the annual seasons. A good example is that of the green alga *Volvox* (fig. 11). I have previously argued that *Volvox*'s (and other small algaes) asexual-sexual

(a)　1 cm

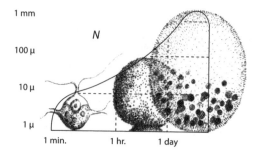

(b)　1 cm

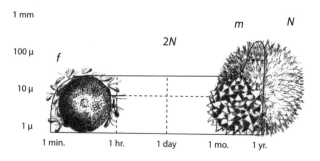

Fig. 11. The drawing at the top is the asexual cycle of *Volvox*. One enlarged cell undergoes a series of cell divisions that produce a number of daughter colonies inside the mother colony that are eventually liberated. During the summer this cycle is repeated many times, but in the fall (below) eggs and sperm cells are formed and after fertilization and numerous cell divisions (including meiosis) a resistant body is formed that can survive the rigors of winter. The figures are on a log-log scale. (From Bonner 1965; drawings by Patricia Collins)

cycles are synchronized with the seasons: with the oncoming of winter (or any condition unfavorable for growth), a sexual generation occurs with all the advantages of recombined offspring.[39] With the return of clement conditions, those zygotes that are blessed with a genome most suited for the new environment will go through a period of rapid growth and bloom with repeated asexual generations. It is common for the zygote to be encased in a resistant shell to carry it through the hard times.

Note that this alternation of generations also occurs in aphids. This process is possible because they are so small and have short generation times. In the growing season they reproduce asexually by parthenogenesis, and, like *Volvox*, they also have their sexual stage in advance of hard times to come. Again, in aphids, because they are small and have the benefit of sexuality, which is switched on at a time when genetic diversity is selectively advantageous, and off (by parthenogenesis) when there is the opportunity for a spurt of rapid multiplication in a steady, benign environment. A control system allows the period of randomness to be turned on or off depending on the season.

It is important to emphasize that this difference in the reproduction of large and small organisms is entirely related to their size. The generation times of large organisms are of sufficiently long duration

---

[39] Bonner 1958.

to span a number of annual seasons; they cannot fit numerous asexual generations into one annual cycle of climate. They have no way of taking advantage of a short annual growing season with rapid asexual reproduction.

Some notable exceptions occur among larger organisms such as vascular plants where even large trees, like aspen, reproduce both sexually and asexually. The two sexes in the form of ovules and pollen produce seeds in the normal fashion and benefit from all the variation-control mechanisms of meiosis, involving the reshuffling of the genes and incorporating new mutations ready to be culled by natural selection. But such a tree will send out runners that, as they extend, give rise to offspring trees that are genetically identical to the parent: they form a clone. The process is repeated, and ultimately it will produce a stand of totally related trees. This is a case where growth need not be limited by annual seasons; trees' reproductive stratagems can span a long period of time. Another example is found among animals: some species of lizards produce young without mating. The eggs are activated without fertilization, by parthenogenesis. But these examples are exceptions; mostly the alternation of sexual and asexual cycles is found among small organisms whose generations are synchronized with the seasons.

The control here is twofold. First, there is control of the extent of variation achieved by sexuality. Second, there is taking advantage of ideal growing conditions by switching to asexual reproduction, which allows many rapid generations to follow in succession before the prison of winter sets in. This control of the mode of reproduction is something that is prevalent in small organisms and rare in large ones. So again, size is a key player in determining whether or not there is to be an alternation of sexual and asexual cycles in the life history of an organism. The reason is largely mechanical: the growth of a large organism is so slow and extended that it cannot take advantage of the short seasons, for it spans many of them. The exception is found among the incredibly slow growing trees such as the aspen, where time, greatly extended, allows the rule to be ignored.

In balance, the vast majority of the examples show that with increase in size, the ability to have an asexual cycle vanishes, and with it vanishes the ability to have a strictly nonrandom phase. As with all generalizations in biology, there are exceptions, as we saw in aspens and in some lizards. Aside from these rare reversals, there is a clear general trend towards a decrease in randomness in larger organisms.

There is another quite different possible reversal where larger organisms increase, rather than decrease, their relative ability to produce randomness. It is connected to Sewall Wright's genetic drift, a phenomenon whose importance has been emphasized by Michael Lynch.[40] For drift to lead to a genetic change in a population, the population size needs to be small. It is a well-known fact that

[40] Lynch 2007a, b.

the population density of an organism is inversely related to its size. For instance, in Africa there are many fewer elephants than small rodents per square mile, and there are infinitely more microorganisms than rodents in the same space. This contradicts my argument that random changes are more abundant in small organisms than in large ones. However, the generation of novel gene arrangements in a population that result from drift is no doubt very small compared to the generation of novelty by mutation and the absence or reduction of selection found in small organisms.

CHAPTER 6

The Division of Labor

Two Cases of the Return of Randomness in Higher Forms

The division of labor has arisen a number of times during the course of evolution, and it is determined in different ways. First there is the conventional method associated with organisms that develop from a single cell, such as an egg that undergoes repeated cleavages with the increase in size. Then there are those cases where the division of labor arises in separate units, be they cells, as in cellular slime molds, or whole organisms, as in insect societies. What will be novel here is that in these latter cases there can be specially engineered periods of nongenetic or phenotypic variation that play a key role in determining the division of labor. It is a return to randomness—where randomness is put to good use.

Both cases involve aggregations (of cells or individual insects) that are characteristic of some nonaquatic, terrestrial organisms. This is true for cellular slime molds, where the variation leading to a division of labor can be in the size and the age of the cells; and similarly in social insects, where also the size or the age of the individual workers may determine the division of labor. The range of those nongenetic periods—their upper and lower limits—is under genetic control and has arisen through natural selection. Organisms that become larger by aggregation put this random variation to work as an essential part of their development—a phenomenon not generally appreciated.

Development by Successive Cleavages

As background, first let us consider conventional development. In the development of most animals and plants, multicellularity is achieved by the cleavage of a single cell, either a fertilized egg or an asexual reproductive cell, and the cleaving cells adhere to one another. This method of size increase first arose in our aquatic ancestors and is retained in aquatic as well as terrestrial descendants. During the course of such a development of successive divisions, some of the cells acquire special characteristics; they are the beginnings of a division of labor.

The timing of such events is under strict genetic control. For instance, the factors that are responsible for the emergence of different cell types may become spatially distributed in the fertilized egg before cleavage begins, so that the descendant cells, by successive cleavages, will contain different factors. This is classical mosaic development. On the other hand, in many organisms all the cells of an embryo retain most of the factors that were present in the egg, so that if in such cases a portion of the embryo is separated, it will produce a complete, diminutive individual. This is regulative development. The division of labor is laid down at a later time in development and is achieved by local differences in the environment of a cell that cause it to differentiate into a particular cell type. The appreciation of these facts is ancient and goes back into the nineteenth century. Note that the main difference between mosaic and regulative development is a matter of timing.

Let us now give an example of how the division of labor arose in a primitive aquatic organism whose increasing size during development is achieved by a series of cell divisions and where the daughter cells stick to one another. In general, this is the case for multicellular animals and plants, all of whom descended from aquatic ancestors.

Volvox and Its Relatives

Volvox is an ancient and primitive alga that has a minimum division of labor, and therefore it is ideally suited to examine how its two cell types evolved.

There is good molecular phylogenetic evidence that *Volvox* and its smaller relatives arose from a unicellular ancestor.[41] The present-day volvocine species come in a range of sizes. The larger species have two cell types (vegetative and reproductive), while the smaller ones have only one (all the cells are first vegetative, and then they all become reproductive). The particular species that exist today are not a direct evolutionary sequence of size increase as used to be thought, for smaller species may have large ancestors, and vice versa. Consider them as occupants of size niches that might have a smaller or a larger immediate ancestor. In other words, the correlation between size and division of labor is not determined by inheritance, by ancestry, but directly by size.

In the larger volvocine algae the division of labor involves a separation of the reproductive cells—either sexual or asexual—and the somatic cells that specialize in photosynthesis and locomotion. In asexual reproduction a few cells in the southern hemisphere of the growing colony undergo an asymmetrical division. The larger of the daughter cells fails to produce flagella and becomes progressively larger (the gonidium). It goes through a series of cleavages and produces a daughter colony within the mother colony.

It has been known for some time that by mutation a large *Volvox* can revert to the behavior of a small species in which all the vegetative cells turn to

[41] Kirk1998; Herron and Michod 2007.

the reproductive mode and become minute daughter colonies.[42] The evolutionary shift from one cell type to two is a relatively minor step. There is good evidence that the first step to this division of labor may well have been a single mutation that controls the arrest of cell division in some cells to produce the larger reproductive cells. And this step was no doubt selectively advantageous to such a large colony.[43] For organisms that become larger by a sequence of cell divisions, which is true of most organisms, this example from *Volvox* and its relatives gives us insight into how the first step towards a division of labor might have occurred.

Cell Societies

Cell size in unicellular organisms may vary in a random spread within a fixed range, as was first observed by H. S. Jennings working with protozoa. J. E. Ackert has provided a model set of experiments to illustrate the point.[44] He started with an average size individual of the ciliate *Paramecium*, whose length could be accurately measured, and cloned it to produce 110 individuals through asexual division. The progeny varied in size, and he selected and cloned both the smallest individual and the largest. Their

[42] Review: Kirk 1998.

[43] For a good review of the possible advantages of size in the volvocine algae see Kirk 1988, pp. 54–67. There are a number of possibilities, the most obvious being that large size protects the individual from being captured by filter feeders such as rotifers.

[44] Ackert 1916.

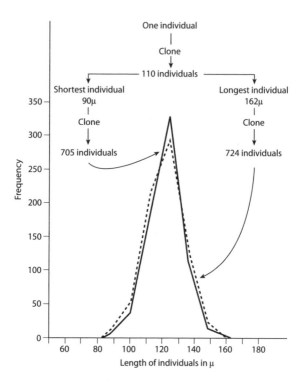

Fig. 12. Range variation in *Paramecium*. The smallest and the largest individual when isolated produce clones with the same size range. (From Ackert 1916)

respective progeny were also measured and showed not only an identical size frequency distribution with the initial parent clone, but with one another as well (fig. 12). In other words, the individual cell size is not genetically determined, although there is every reason to believe the mean size, or the size range, is. For this reason I have called this kind on phenotypic variation *range variation*.[45] The variation within that range is entirely nongenetic. This nongenetic variation was also shown in some ancient experiments of Johannsen,[46] who sorted beans in the same way. If the beans came from a pure, inbred strain, both the smallest and the largest beans produced the same range of sizes consistently for a number of generations (fig. 13). However, with genetically heterogeneous strains one could select for larger or smaller offspring. Note that the *Paramecia* of Ackert are also genetically identical—they are clones.

One wonders if this property of microbial range variation is ever put to some use, or is it no more than an anomalous curiosity. In microorganisms I can think of two cases where range variation plays a positive role from the point of view of natural selection, one of which involves a ciliate, and the other—a much more compelling case—is found in the social amoebae.

Some years ago, A. C. Giese showed that if one starved the ciliate *Blepharisma*, the colony became dimorphic: big cells and small ones.[47] There is nothing to suggest that this size difference is genetically determined because both of the extreme sizes, if given the right amount of food, would return to the normal size. The dimorphic-size ciliates behaved very curiously: the larger individuals ate the smaller ones. Starvation reduces them to cannibalism, and

[45] Bonner 1965.
[46] Johannsen 1911.
[47] Giese 1938.

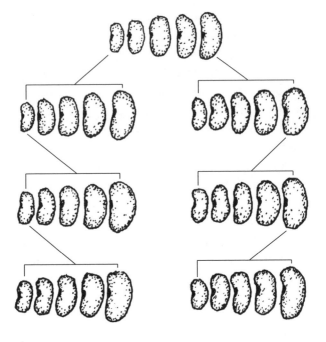

Fig. 13. Pure lines (genetically similar) of beans showing range variation. With repeated selection the smallest and the largest produce the same span of offspring sizes. (After Johannsen 1911)

the size variation makes it possible. In some way, starvation causes the dimorphism with the result that they have a strategy to delay extinction. It is rather like a small boat full of starving, shipwrecked sailors, who have been known to follow a similar survival strategy. How starvation produces the size bimodality is not known, but it is likely to have a simple answer. We have here a case where nongenetic variation produces a way of slowing the extinction of a starving population. Clearly this is evolutionarily advantageous, yet is achieved stochastically. The ability to manage this behavior is no doubt genetically determined, but the specific behavior itself is not. This is a case where random, nongenetic variation can be useful and produce a division of labor.

My second example in which range variation is put to use is seen in the cellular slime molds. Most species have two cell types: stalk cells and spores. The origin of a two-cell-type state must have quite a different history from that of *Blepharisma*. There are some species of slime molds that have only one cell type: in members of the genus *Acytostelium*, all the amoebae first secrete a noncellular stalk and then all turn into spores. It had always been assumed that *Acytostelium* was ancestral, but from recent molecular phylogenetic studies it is now known they are most likely descendants of species that possessed both stalk cells and spores.[48] This leads to the possibility that the same principle applies here that we saw in *Volvox* and in other volvocine algae. *Acytostelium* species are all relatively small compared to most other cellular slime molds, and this may be the reason they have evolved a single cell type unlike their two-cell-type ancestors. (Some of those two-cell-type species are also very small, but this may be secondarily derived.) In both cellular slime molds and

[48] Schaap et al. 2006; Romeralo et al. 2011.

the volvocine algae, size appears to play a key role in the division of labor.

In the most studied species, *Dictyostelium discoideum*, the early signs of the two cell types appear in the migrating slugs: they are the prespore and prestalk zones (fig. 7). All the anterior cells form the stalk and lift the mass of spores up into the air. In doing so they swell, develop large vacuoles and sturdy cellulose cell walls, and in this process perish. They are the ultimate altruists. Regardless of the size of the cell mass, they maintain rough proportions: approximately 20 to 30 percent of the amoebae in the anterior of the slug become stalk cells, and the remaining posterior amoebae turn into spores. Before I discuss the regulation of the proportions, let us first examine the prime question of what determines the cell fate: how is it decided whether an amoeba becomes a stalk cell or a spore?

It turns out that the amoebae are not all the same. As migration takes place, some of the amoebae rush to the front of the slug and the others lag behind. And those active ones that move forward are rushing to their death, for they form the stalk. The really interesting discovery made by a number of workers is that the differences among the amoebae is how well fed they are: the lean amoebae are the active ones that rush forward and become stalk cells, while the well-fed ones, the laggards, become the spores.[49] In any population of feeding amoebae, when the

---

[49] Review: Bonner 2009.

bacterial food supply is gone there will be the gamut from replete amoebae to virtually starved ones. This is because the amoebae that have just divided will have less food inside them than those that are about to divide. So in this case, the age of an amoeba and the amount of food it contains are directly corre-lated. As soon as they gather together by aggrega-tion to form a slug, the lean, younger amoebae move forward and leave the older, well-fed ones behind. It should be emphasized that the degree of satiation is a continuum; there is no sharp division line between the haves and the have-nots; all that counts is "more than" or "less than." And one more point should be added to stress that the differences in the amoebae are a relative matter, for if one cuts off a segment— any segment—of a slug it will regulate and produce normal stalk cells and spores. In other words, the fate of a cell is not fixed: an amoeba that might be destined to become a stalk cell could also become a spore, and vice versa.

To return to the question of how two cell types evolved in these slime molds, it is clear that, unlike *Volvox*, it is not due to some simple mutation that affects cell divisions, but is based on the phenotypic differences of the amoebae—on how well fed they are. More specifically, they exhibit range variation, and they put this nongenetic variation to good use. One could even argue that this a particularly ef-ficient way of assigning cell types, for the anterior cells are going to make the stalk require only enough energy to do so before they die, while the posterior cells, by having a store of food, will help them when they germinate and seek their first meal of bacteria.

So here is a case where nongenetic variation provides an effective way of allotting the differentiation of two cell types in a primitive organism. It must be understood that the whole process is within a genetic framework. The ability to be able to make a spore or a stalk cell is in itself under orthodox genetic control. One of the more important of these controls is the mechanism for keeping the spore-stalk ratio roughly constant in cell masses of different sizes. There is evidence that the posterior prespore zone secretes a key molecule (or molecules) that dominates the anterior amoebae and forces them to sacrifice themselves for the good of the selfish spores. But remember it is not so much of a sacrifice because they share the same genes.[50]

There is no evidence that the cellular slime molds are the ancestors of any other organisms, ancient though they be. Nor is there any evidence that eventually genes took over the responsibility of designating which amoebae will become spores and which stalk cells, as they did right from the beginning in *Volvox*. So in some respects they are an evolutionary dead end, yet a highly successful one. Despite their early origin in Earth history, today they exist in great abundance in the soil all over the globe, from the arctic tundra to the lush tropics. It is often claimed

[50] Ibid.

that for total tonnage, ants and other social insects reign supreme, but I wonder if they might be topped by social amoebae, who would in turn be topped by bacteria.

Insect Societies

This way of determining the division of labor in slime molds is characteristic of those forms that initially became multicellular on land. It involves the component units—cells—that are initially separate and come together by aggregation. As we will see, the division of labor in social insects, where the individual insects also form aggregations (although of a very different nature), uses range variation as well. This is seen in the division of labor among the neuter castes in ants, bees, wasps, and termites.

The question of how different castes divide the labor is an ancient concern, and early work favored the idea that caste determination is nongenetic. The prime influence is the amount of food provided the larvae: for instance, in ants a minimal diet produces a small worker, and the smallest workers in the colonies are confined to specific tasks, such as helping the egg-making machine that is the queen, or taking on general household duties and keeping the nest clean. The larger workers, who have received more food as larvae, may be the foragers who go out of the nest to gather food. A perfect example of this is also found among bumblebees. Their nests are rather small and not highly organized, and the brood cells at the center receive much more attention and are fed more than the cells at the periphery. As a result there can be a great difference in the size of the emerging worker.[51]

It is not only how much the larvae are fed, but also what they are fed. In honey bees special large cells are built to spawn a new queen. Along with more food, the food is supplemented with rich royal jelly that is secreted from the glands of the workers and is fed directly to the royal larva by the workers.

An important point should be made here. These nutritional ways of determining how the labor is divided parallels what we found in the spore-stalk cell determination in slime molds; both are examples of range variation, and the variation within the range is based on the amount of food each individual receives.

There is another parallel between social insects and slime molds. In the latter we saw special substances produced by the presumptive spores that pushed the anterior cells to become stalk cells, and this presumably is the way in which a stable spore-stalk ratio is controlled regardless of the total number of cells. In social insects the proportions of the different castes are controlled in a similar way, as is seen particularly clearly in termites, where each individual juvenile undergoes a series of molts and can switch from one caste to another over time. If

[51] Couvillon and Dornhaus 2009.

there are fewer than the normal number of soldiers compared to workers in a colony, the developing individuals will start producing more soldiers until the normal balanced is reached. This is achieved by a decrease in a chemical that is passed from individual to individual and inhibits soldier development. The decrease is simply the result of there being fewer workers to produce it.

Caste is determined by nongenetic means, yet let us consider this statement more carefully. It may be true for the stimulus, but the capacity to respond to these external cues is of course genetically based. It may not be true for simple size differences determined by the amount of food, but consider the capacity to grow huge mandibles in ants or to produce the toxic spray apparatus of termite nasute soldiers. These striking morphologies clearly reflect a genetically determined response system.

Now we come to a particularly interesting—and somewhat different—example. In honeybees, genes have managed to infiltrate into the nongenetic control of the division of labor. The classical view is that the different tasks of worker honeybees was determined by their age.[52] The younger workers are confined to nest duties, and the older bees become foragers. This is a temporal range variation that is well established.

Another surprising element has been appreciated recently in the key discovery of Robert Page and

[52] Lindauer 1961.

Gene Robinson.[53] Normally the queen is fertilized by many males (roughly from seven to seventeen) and she stores all these genetically diverse sperm in her reproductive tract. As a result, her worker offspring are not genetically identical. By careful supervision colonies were raised with a queen fertilized by only one male. The resulting colonies showed a tendency to produce either a disproportionate number of workers favoring a particular task such as housekeeping or foraging. The foragers show a preference for gathering nectar and pollen. In other words, the behavior of the worker bees has a genetic component. That such a state might become selectively advantageous was demonstrated by some observations of Matilla and Seeley, who showed that if they compare a normally established hive with one in which the queen has been inseminated by only one male, the normal, multi-male colony will harvest a far greater amount of honey and produce more offspring than the single-male colony.[54] So natural selection will favor polyandry if genes biasing a task preference were to creep in.

To me this is an unexpected and surprising phenomenon. The age method of determining the labor seems superior, and the genetic control involving polyandry is cumbersome by comparison. Perhaps the answer lies in the very nature of natural selection where the genetic control of a phenomenon will creep in with the slightest encouragement. This ge-

[53] Page and Robinson 1991.
[54] Matilla and Seeley 2008.

netic system of producing a division of labor has not supplanted the nongenetic, worker-age system: the two work together. The nongenetic method of dividing the labor is totally sufficient in itself, but here genes and selection invade even though they appear to be superfluous—they are not needed. Yet they have devised a way to be involved. It shows that even though randomness is doing a splendid job, natural selection nevertheless appears. It is almost as though natural selection is at war with randomness: while randomness is ubiquitous, natural selection is still the ultimate master.

The division of labor among living organisms has arisen independently a number of times. It has been associated with size increase of organisms that grow by successive cleavages, which is the case for all multicellular animals and plants, all of which are descendants of distant aquatic ancestors. It also has arisen more than once in a different form in both cell and animal societies. In general, it has arisen through the natural selection of genetically controlled events, but, as we have seen, in cell and animal societies nongenetic, random mechanisms have been devised as effective ways of dividing the labor.

CHAPTER 7

Envoi

The purpose of this essay has been to give a balanced view of evolution by showing how big a role randomness plays. I think randomness is necessary to counteract the tremendous power of natural selection that to some degree blinds our vision. Selection is the supreme mechanism that brings order out of chaos, and for that reason it is quite rightly foremost in our minds in all matters concerning evolution. It is for this reason that randomness is often ignored and sometimes rejected because of our natural selection mind-set. But, as I have pointed out, there could be no natural selection without randomness, for it is the foundation upon which natural selection is built. It provides the fodder for selection; without it everything would be the same and no change, no evolution would be possible. Randomness keeps appearing in different ways and at different times during the course of evolution. For many biologists, randomness is the skeleton that can be kept in the closet, and what I have done is bring it out.

For me, one of the most interesting aspects of biological randomness is the effect size has on it. In particular, with increase in size there is an increase in the extent and duration of development. This means a vast increase in the number of gene-controlled, sequential biochemical steps, any one of which could be affected by a random, deleterious mutation, and the result would be the death of the embryo. This internal selection becomes increasingly important the larger the embryo; the greater the number of developmental steps, the greater the opportunities for stifling errors. Development must be an almost perfect machine if a viable individual is to be produced. At the terminal end of all these steps, mutations will no longer be culled by internal selection and will be part of the adult living organism. Now the animal or plant enters the outside world and can be reached by natural selection. It has switched from internal to external (natural) selection.

In a crude sort of way, small organisms are more likely to be involved in randomness than large ones. Because internal selection will play little or no role in their short development, they will produce adult morphological variants in large numbers, and there will be an increased chance that some of them are untouched by natural selection. The big problem is that their neutrality cannot be proved, a difficulty that ignites the passions of many committed adaptationists.

To make a grand summary of some of the major points in this essay, here is a very crude diagram that brings those points together (fig. 14).

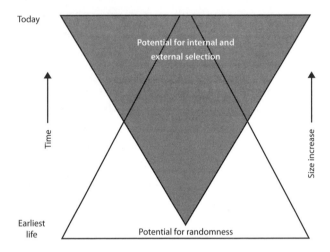

Fig. 14. A crude and simplified diagram reflecting the main themes in this essay. In early life, when organisms were small and relatively simple, random changes were common and natural selection played a less prominent role. Today, with the progressive increase in the size and complexity of organisms, there is a more prominent role of internal selection during development, and natural selection of the mature animals and plants. (I am indebted to Hannah Bonner and Jonathan Bonner for their suggestions for this figure.)

ACKNOWLEDGMENTS

Largely because some find what I have written controversial, I have received more generous help from more kind people than any of my earlier books. I have been berated for straying from the beaten path, or encouraged because I had done so. During the course of all these critiques I was forced to become increasingly clear and precise; it was because some of the ideas were new to me as well that the road to getting it right was very bumpy. And it would not have been possible without the help of my stalwart readers.

The main bone of contention was that I seemed to be slighting the accepted view that organisms were always the product of natural selection and that all their features were adaptations, proven or assumed. The idea that there are exceptions in the case of small organisms seemed heretical, which stimulated me to try to show that this was not the case.

I have been working on this project more or less nonstop for six years. In the beginning the person who gave me the initial encouragement (along with penetrating critiques) and continued urging me on all along the way is Vidyanand Nanjundiah. The other key person is Mary Jane West-Eberhard, who began helping me at an early stage and has been my severest critic, but at the same time also my most helpful one. She not only purified my thoughts, but made clear to me what the rest of the evolutionary world was thinking. I owe a big debt to both of these friends.

I want to give special thanks to others, some of whom made points along the same lines. They are as follows: my A1 critics for many years, my colleagues Henry Horn and Ted Cox; and Tom Doak, whose conversations with me at an early stage were pivotal; Tom was the first to make me realize that my thoughts were not going to be accepted unchallenged.

There are others who played a big and helpful role along the way for which I am most grateful: Leo Buss, Peter and Rosemary Grant, Alison Kalett, Laura Katz, Egbert Leigh, Dan McShea, Jonathan Payne, and Jonathan Weiner. And finally I would like to thank Slawa Lamont, whose suggestions have been so helpful and whose support, when the going was not exactly smooth, sustained me.

The publishing process with the Princeton University Press has been singularly pleasant because of the great help I received along the way: my editor, Alison Kalett, who provided wisdom and good judgment that opened the door for publication; Alice Calaprice, who did her copyediting wizardry on my prose; Pamela Schnitter, who designed the book, much to the pleasure of the author and the reader; and Debbie Tegarden, who, with loving care, eased the preparation of the book through all the details of actually making a book.

BIBLIOGRAPHY

Ackert, J. E. (1916) On the effect of selection in *Paramecium. Genetics* 1:387–405.

Alvarez-Curto, E., D. E. Rozenm, A. V. Ritche, C. Fouquet, S. L. Baldauf, and P. Schaap (2005) Evolutionary origin of cAMP-based chemoattraction in the social amoebae. *Proc. Nat. Acad. Sci. USA* 102: 6385–6390.

Asghar, A., M. Groth, O. Siol, F. Gaube, C. Enzensperger, G. Glöckner, and T. Winckler (2012) Developmental gene regulation by an ancient intercellular communication system in social amoebae. *Protist* 163: 25–37.

Bonner, J. T. (1958) The relation of spore formation to recombination. *Amer. Nat.* 92: 193–200.

Bonner, J. T. (2000) *First Signals: The Evolution of Multicellular Development.* Princeton: Princeton University Press.

Bonner, J. T. (2006) *Why Size Matters.* Princeton: Princeton University Press.

Bonner, J. T. (2009) *The Social Amoebae.* Princeton: Princeton University Press.

Borass, M. E., D. B. Seale, and J. E. Boxhorn (1998) Phagotrophy by a flagellate selects for colony prey: A possible origin of multicellularity. *Evolutionary Ecology* 12: 153–164.

Couvillon, M. J., and A. Dornhaus (2009) Location, location, location: Larvae position inside the nest is correlated with adult body size in worker bumble-bees (*Bombus impatiens*). *Proc. R. Soc. Lond. B.* 276: 2411–2418.

Darwin, C. (1959 *et seq.*) *On the Origin of Species.* London: John Murray.

Davidson, E. H. (2009) Network design principles from the sea urchin embryo. *Curr. Opin. in Genetics Dev.* 19: 535–540.

De Beer, G. (1962) *Embryos and Ancestors* (3rd ed). Oxford: Clarendon Press.

Finch, C. E., and T.B.L. Kirkwood (1999) *Chance, Development, and Aging.* New York: Oxford University Press.

Finlay, B. J., and T. Fenchel (2004) Cosmopolitan meta-populations in free-living microbial eukaryotes. *Protist* 155: 237–244.

Giese, A. C. (1938) Cannibalism and gigantism in *Blepharisma. Trans. Amer. Micr. Soc.* 57: 245–255.

Gould, S. J., and R. C. Lewontin (1979) The Spandrels of San Marco and the Panglossian paradigm: A critique of the adaptionist programme. *Proc. R. Soc. Lond. B* 205: 581–598.

Grant, P. R., and B. R. Grant (2011) *How and Why Species Multiply: The Radiation of Darwin's Finches.* Princeton: Princeton University Press.

Grant, V. (1977) *Organismic Evolution.* San Francisco: W. H. Freeman & Co.

Herron, M. D., and R. E. Michod (2008) Evolution of complexity in the volvocine algae: Transitions in individuality through Darwin's eye. *Evolution* 62-2: 436–451.

Horn, E. G. (1971) Food competition among cellular slime molds (Acrasiales). *Ecology* 52: 475–484.

Johannsen, W. (1911) The genotype conception of heredity. *Amer. Nat.* 45: 129–531.

Katz, L. A. , G. B. McManus, O.L.O. Snoeyenbos-West, A. Griffin, K. Pirog, B. Costas, and W. Foissner (2005) Reframing the "Everything is everywhere" debate: Evidence for high gene flow and diversity in ciliate morphospecies. *Aquatic Microbial Ecol.* 41:55–65.

Kimura, M. (1983) *The Neutral Theory of Molecular Evolution.* Cambridge, U.K.: Cambridge University Press.

Kirk, D. L. (1998) *Volvox.* Cambridge, U.K.: Cambridge University Press.

Lindauer, M. (1961) *Communication among Social Bees.* Cambridge, Mass.: Harvard University Press.

Lynch, M. (2007a) *The Origins of Genome Architecture.* Sutherland, Mass.: Sinauer.

Lynch, M. (2007b) The frailty of adaptive hypotheses for the origin of organismal complexity. *Proc. Nat. Acad. Sci. USA.*104: 8597–8604.

Matilla, H. R., and T. D. Seeley (2007) Genetic diversity in honey bee colonies enhances productivity and fitness. *Science* 317: 362–364.

McShea, D. W. (2002) A complexity drain on cells in the evolution of multicellularity. *Evolution* 56: 441–452.

McShea, D. W. (2005) The evolution of complexity without natural selection, a possible large-scale trend of the fourth kind. *Paleobiology* 31(2, Suppl.): 146–156.

McShea, D. W., and R. N. Brandon (2010) *Biology's First Law: The Tendency for Diversity and Complexity to Increase.* Chicago: University of Chicago Press.

Nee, S., and Stone, G. (2003) The end of the beginning for neutral theory. *Trends in Evol. and Ecol.* 18: 433–434.

Newman, S., G. Forgacs, and G. B. Müller (2006) Before programs: The physical origination of multicellular forms. *Internat. J. Developmental Biol.* 50: 289–99.

Page, R., and G. E. Robinson (1991) The genetics of division of labor in honey bee colonies. *Adv. Insect Physiol.* 23: 117–167.

Parent, C. A. and P. N. Devreotes (1996) Molecular genetics of signal transduction in *Dictyostelium. Ann. Rev. Biochem.* 65: 411–440.

Payne, J. L., A. G. Boyer, J. H. Brown, S. Finnegan, M. Kowalewski, R. A. Krause, Jr., S. K. Lyons, C. R. McClain, D. W. McShea, P. M. Novak-Gottshall, F. A. Smith, J. A. Stempien, and S. C. Wang (2009) Two-phase increase in the maximum size of life over 3.3 billion years reflects biological innovation and environmental opportunity. *Proc. Nat. Acad. Sci. USA* 106: 24–27.

Romeralo, M., J. C. Cavender, J. C. Landolt, S. L. Stevenson, and S. L. Baldauf (2011). An expanded phylogeny of social amoebas (Dictyostelia) shows increasing diversity and new morphological patterns. *BMC Evolutionary Biology* 11: 84.

Satish, N., S. Sultana, and V. Nanjundiah (2007) Diversity of soil fungi in tropical deciduous forest in Mudumalai, southern India. *Current Science* 93: 669–677.

Schaap, P., T. Winkler, M. Nelson, E. Alvarez-Curto, B. Elgie, H. Hagiwara, J. Cavender, A. Milano-Curto, D. E. Rozen, T. Dingermann, R. Mutzel, and S. L. Baldauf (2006) Molecular phylogeny and evolution of morphology in the social amoebas. *Science* 314: 661–663.

Thompson, D'A. W. (1917 et seq.) *On Growth and Form*. Cambridge, U.K.: Cambridge University Press.

Valentine, J. W., A. G. Collins, and C. P. Meyer (1994) Morphological complexity increase in metazoans. *Paleobiology* 20: 131–142.

Waddington, C. H. (1957) *The Strategy of the Genes*. London: George Allen & Unwin.

Whyte, L. L. (1965) *Internal Factors in Evolution*. New York: G. Braziller.

INDEX